# Business Analytics Using SAS® Enterprise Guide® and SAS® Enterprise Miner™
## A Beginner's Guide

Olivia Parr-Rud

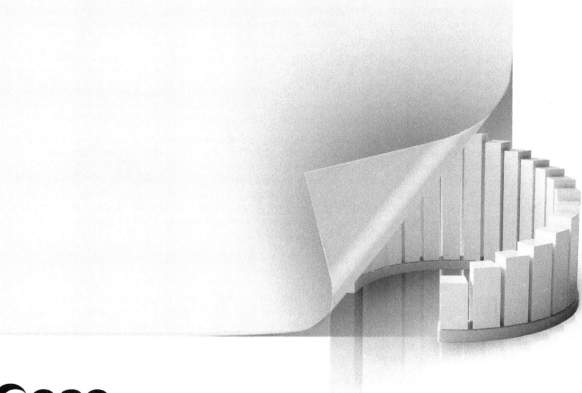

**support.sas.com/bookstore**

The correct bibliographic citation for this manual is as follows: Parr-Rud, Olivia. 2014. *Business Analytics Using SAS® Enterprise Guide® and SAS® Enterprise Miner®: A Beginner's Guide*. Cary, NC: SAS Institute Inc.

**Business Analytics Using SAS® Enterprise Guide® and SAS® Enterprise Miner®: A Beginner's Guide**

SAS provides a complete selection of books and electronic products to help customers use SAS® software to its fullest potential. For more information about our offerings, visit **support.sas.com/bookstore** or call 1-800-727-3228.

# Contents

# About This Book

## Purpose

This book serves as a tutorial for data analysts who are new to SAS Enterprise Guide and SAS Enterprise Miner. It provides valuable hands-on experience using powerful statistical software to complete the kinds of business analytics common to most industries. With clear, illustrated, step-by-step instructions, it will lead you through examples based on business case studies. You will formulate the business objective, manage the data, and perform analyses that you can use to optimize marketing, risk, and customer relationship management, as well as business processes and human resources.

## Prerequisites

If you are a savvy business person with a desire to understand what drives your business, then this book can help you get started. You need access to SAS Enterprise Guide or SAS Enterprise Miner software; we provide you with example data to get started, but you will need data to analyze. An understanding of basic statistics is helpful, but not required.

## Organization

The book begins by helping you determine and structure the objective of your analysis in accordance with the goals and objectives of your organization or department.

Chapter 2 describes types and sources of data for analysis. Chapter 3 offers an overview of common business analyses, covering both descriptive and predictive analysis. Chapter 4 shows you how to construct a data set for analysis. Chapter 5 details step-by-step instructions for a simple descriptive analysis. Chapter 6 offers the same level of detail for a typical market analysis. Chapters 7 and 8 offer a step-by-step guide to cluster and tree analyses, respectively. Each chapter concludes with a section headed "Notes from the Field," which offers related business advice and leadership tips.

To conclude, Chapter 9 brings several concepts together in a full step-by-step case study for building and comparing predictive models, culminating in final "Notes from the Field."

# Examples

SAS Institute and SAS Press provide access to software updates and the author's example data sets so that you can practice the examples in this book.

## Software Used

The software packages used in the writing of this book are SAS Enterprise Guide 6.1 M1 and SAS Enterprise Miner 13.1. Although these are the latest versions available at the time of publication, new features will appear in later releases. Visit the SAS Products and Solutions webpage for updates and enhancements to all SAS system software at http://www.sas.com/en_us/software/all-products.html.

## Data Sets

You can access the data used in the author's examples by linking to this book's author page at http://support.sas.com/publishing/authors. Select the name of the author. Then look for the cover thumbnail of this book, and select *Example Data* to display the SAS data sets associated with this book.

For an alphabetical listing of all books for which example code and data sets are available, see http://support.sas.com/bookcode. To display a book's example code, select its book title.

If you are unable to access data sets through the website, email saspress@sas.com.

# Additional Help

Although this book illustrates many analyses regularly performed in businesses across industries, questions specific to your aims and issues may arise. To fully support you, SAS Institute and SAS Press offer you the following help resources:

- For questions about topics covered in this book, contact the author through SAS Press:

  - Send questions by email to saspress@sas.com; include the book title in your correspondence.

  - Submit feedback on the author's page at http://support.sas.com/author_feedback.

- For questions about topics in or beyond the scope of this book, post queries to the relevant SAS Support Communities at https://communities.sas.com/welcome.

- SAS Institute maintains a comprehensive website with up-to-date information. One page that is particularly useful to both the novice and the seasoned SAS user is its Knowledge Base.

Search for relevant notes in the "Samples and SAS Notes" section of the Knowledge Base at http://support.sas.com/resources.

- Registered SAS users or their organizations can access SAS Customer Support at http://support.sas.com. Here you can pose specific questions to SAS Customer Support; under *Support*, click *Submit a Problem*. You will need to provide an email address to which replies can be sent, identify your organization, and provide a customer site number or license information. This information can be found in your SAS logs.

## Keep in Touch

We look forward to hearing from you. We welcome questions, comments, and concerns. If you want to contact us about a specific book, please include the book title in your correspondence. For a complete list of books available through SAS, visit http://support.sas.com/bookstore.

## SAS Books

Reach our bookstore by phone, fax, or email:

- Phone: 1-800-727-3228

- Fax: 1-919-677-8166

- Email: sasbook@sas.com

## SAS Book Report

Receive up-to-date information about all new SAS publications via email by subscribing to the SAS Book Report monthly eNewsletter. Visit http://support.sas.com/sbr.

x

# About the Author

**Olivia Parr-Rud**, an internationally recognized expert in predictive analytics, business intelligence, and innovative leadership, founded the SAS Data Mining Users Group, having been a SAS user since 1991 and a SAS instructor and conference presenter for many years. She hosts the popular VoiceAmerica Business radio show *Quantum Business Insights* (http://www.voiceamerica.com/show/2240/quantum-business-insights) and is a thought leader in the integration of analytic tools and leadership practices to optimize performance and organizational agility. Her predictive analytics research founded her first book, *Data Mining Cookbook: Modeling for Marketing, Risk and Customer Relationship Management* (Wiley 2000), which unveiled links between the global economy and organizational dynamics, a topic featured in her second book, *Business Intelligence Success Factors: Tools for Aligning Your Business in a Global Economy* (Wiley and SAS Institute Inc. 2009). Her current research will develop a model that tracks the alignment of business intelligence with human intelligence. Parr-Rud holds a B.A. in mathematics and an M.S. in statistics. In addition to public speaking, she offers training and consulting in both predictive analytics and business leadership, with clients that include numerous Fortune 500 companies. For more, visit http://oliviagroup.com.

Learn more about this author by visiting her author page at http://support.sas.com/publishing/authors/parr-rud.html. There you can download free book excerpts, access example code and data, read the latest reviews, get updates, and more.

# Chapter 1: Defining the Business Objective

## Introduction

In today's highly competitive global economy, organizations are increasingly dependent on their ability to leverage accurate, accessible, actionable data. This increased dependence on data requires a new kind of visionary leadership focused on a smart investment in an enterprise data management system, the right analytic talent to leverage it, and a culture designed to support "big data."

A culture of big data includes powerful analytic tools. Advanced analytics and reporting software platforms, such as SAS Enterprise Guide and SAS Enterprise Miner, enable the user to extract deep insights from big data and its underlying patterns. This book is designed to guide the business analyst or manager as he or she seeks these insights by providing case studies and everyday examples.

In this chapter, you will learn the importance of clearly defining the objective of any data analysis project.

## Setting Goals

The use of data analysis is now commonplace in business across industries. Many applications, such as customer profiles or response or risk models, are quite straightforward. However, as companies attempt to develop models that address more complicated measures such as customer retention and lifetime value, the importance of clear goals is magnified. Failure to define the goal correctly can result in wasted dollars and a lost opportunity.

The first and most important step in any data analysis project is to establish a clear goal, not a goal defined only by the data or the method, but a goal that makes sense to the business as a whole. In other words, the goal of the analysis must be defined in terms of how it will help the business reach its strategic goals. When you engage the stakeholders, questions and analytic methods will likely include the following:

- *Do you need to understand the characteristics of your current customers?* This goal may involve looking for averages or segmenting your customer base and creating profiles.
- *Do you need to attract new customers?* Response modeling on customer acquisition campaigns enables you to lower the marketing costs of attracting additional customers.
- *Do you need to avoid high-risk customers?* Risk or approval models identify customers or prospects that have a high likelihood of incurring a loss for the company. In financial services, for example, a typical loss comes from nonpayment on a loan. Insurance losses result when claims filed by the insured outweigh the calculated loss reserves.
- *Do you need to make your unprofitable customers more profitable?* Cross-sell and up-sell target models can be used to increase profits from current customers.
- *Do you need to retain your profitable customers?* Retention, or "churn," models identify customers with a high likelihood of lowering or ceasing their current levels of activity. The identification of these customers before they change their behavior enables the construction of strategies and actions to retain them. The cost of retaining a customer is often less than the cost of winning back a customer.
- *Do you need new customers to be profitable?* A lifetime value model identifies prospects with a high likelihood of being long-term profitable customers. Combining several measures, such as response, activation, and attrition, can help you target prospects that have the highest value over the life of the product or service.
- *Do you need to win back your lost customers?* Win-back models are built to target former customers.
- *Do you need to improve customer satisfaction?* In today's competitive market, customer satisfaction is crucial to success. Combining market research with customer profiling is an effective method of measuring customer satisfaction.
- *Do you need to increase sales?* You can increase sales in several ways. A new-customer-acquisition model can grow the customer base, leading to increased sales. Cross-sell and up-sell models can also be used to increase sales.
- *Do you need to determine what products or services to bundle or offer sequentially?* Affinity analysis can identify products that have a high probability of being purchased together or within a narrow time frame.
- *Do you need to reduce expenses?* Better targeting through the use of models for customer acquisition and customer relationship management can improve the efficiency of your marketing efforts by reducing expenses.

- *Do you need to determine the most effective channel or sequence of channels?* Analysis of purchase behavior can identify and measure which channel or sequence of channels is most efficient.

- *Do you need to optimize the frequency of your offers?* Analysis based on frequency testing can indicate how often to send an offer. This frequency strategy may vary greatly by channel in terms of both cost and results.

- Do you need to de*liver the right message to the right person at the right time through the right channel?* This objective is every marketer's overall goal and encompasses all of the more specific goals.

So far, the list of goals concerns selling a product or service. Other business uses of data analysis that can result in cost reductions and higher profits may involve answering the following related questions:

- *Do you need to avoid process failure?* Companies in a variety of industries are modeling data collected from production lines, customer feedback, hospital error, and similar sources to predict breakdowns and take corrective actions.

- *Do you need to analyze health treatments?* Predictive models in the life sciences have saved money and lives. Models that predict disease outbreaks are used to calculate inventory in high-risk geographic regions.

- *Do you need to optimize your inventory?* Knowing how many of what products to have on hand in a specific location can save time and money.

- *Do you need to optimize your staffing?* Demand can be predicted, which enables stores, resorts, and other businesses to know exactly whom to have on-site or available for work.

- *Do you need to predict the best location for a future store, restaurant, or other business?* Planning growth can be facilitated by the prediction of sales for a specific geographic area.

For numerous other examples of goals, see *Predictive Analytics* (Siegel and Davenport 2013).

To achieve your goals, you will choose appropriately from a variety of analyses. The two major categories of analysis are *Descriptive Analysis* and *Predictive Analysis*.

# Descriptive Analyses

*Descriptive analysis* is a technique that allows you to view and measure your company and customer characteristics. The results of the analysis can be used to guide decisions in every area of your company. Therefore, descriptive analysis is an essential first step to managing your business.

## Customer Profile

Because your best customers typically drive your company's profitability, analyzing your customer base is a good first step. This provides a snapshot of exactly who is buying your products or services.

A customer profile analysis identifies and measures the characteristics of your most profitable customers. Insights gained from this analysis can be used to enhance and customize your marketing activities. Customer profile analysis involves measuring common characteristics within a population of interest. Demographics, such as average age, gender (percent male), marital status (percent married, percent single, and so on), and average length of residence, are typically included in a profile analysis. Other measures might be more business-specific, such as the age of a customer relationship or the average risk level.

Also, if you find distinct characteristics that define a particularly profitable segment, you can match those characteristics to names on other databases or external lists and purchase new names for marketing. Similarly, if you identify the most and least profitable customers, you can focus your retention efforts and offer varying levels of customer service.

## Customer Loyalty

*Loyalty* is an aspect of the customer profile that is measured by the length of the relationship, the amount of spending, or a combination thereof. Unlike customer profile analysis, which is more of a snapshot of your base, customer loyalty analysis looks at your customer base over time. History tells us that, on average, loyal customers bring in more revenue at a lower cost.

## Market Penetration or Wallet Share

*Market penetration analysis* and *wallet share analysis* are techniques for measuring the performance of your customer base in comparison with the performance of the overall market for your industry. These analyses typically look at these measures across selected demographics or other characteristics, or a combination thereof.

Your company's market penetration compares the number of customers in your base to the possible customers in the overall market. Wallet share compares your revenues to those available in the market. It is powerful for determining how much potential lies with a specific segment as defined by a demographic or behavioral characteristic.

Wallet share analysis can be used to prioritize marketing dollars in industries where the customer has a fixed amount to spend with multiple businesses (their "wallet"). You calculate it by taking the amount that the customer is spending with your company and dividing it by the total that they have to spend with you and your competitors.

## Predictive Analyses

*Predictive analysis* uses statistics, machine learning, or data mining to develop models that predict future events on the basis of current or past data. In business, there are numerous models that are commonly used.

## Marketing Models

*Marketing models* are models that are built and used specifically for growing your business.

### Response

A *response model* is typically the first type of target model that a company seeks to develop. If no targeting was done in the past, a response model can provide a significant boost to the efficiency of a marketing campaign by increasing responses or reducing expenses. The objective is to predict who will respond to an offer for a product or service. It can be based on past behavior of a similar population or some logical substitute. For more detail on data for modeling, see Chapter 2.

### Win-Back

A *win-back model* is used to invite former customers to reconsider their relationship to the business. These models can be powerful, because you already have good behavioral data on these former customers. You should consider risk and retention measures when appropriate.

### Activation

An *activation model* predicts whether a prospect will become a customer. It is applicable only in certain industries, most commonly credit card and insurance. For example, for a credit card prospect to become an active customer, the prospect must respond, have their credit approved, and use the account. If the customer never uses the account, he or she actually costs the bank more than a noncustomer. Most credit card banks offer incentives, such as low-rate purchases or balance transfers, to motivate new customers to activate. An insurance prospect can be viewed in much the same way. A prospect can respond and be approved, but if he or she does not pay the initial premium, the policy is never activated.

### Revenue

A *revenue model* predicts the dollar amount of an expected sale. This model is useful for distinguishing low-value from high-value responders or customers. For example, if a prospect becomes a customer after buying a product online, he or she has a specific value based on the amount of that purchase.

### Usage

A *usage model* predicts the amount of use given to a product or service. This model is most applicable to the telecommunications companies that determine their profits by estimating minutes of usage. Life sciences companies predict disease outbreaks in specific geographic locations, on the basis of doctor visits, sales of prescriptions, and over-the-counter drugs.

### Cross-Sell and Up-Sell

A *cross-sell model* is used to predict the probability or value of a current customer's buying a different product or service from the same company. An *up-sell model* predicts the probability or value of a customer's buying more of the same product or service.

Selling additional products or services to current customers is quickly replacing new customer acquisition as one of the easiest ways to increase profits. Testing offer sequences can help to determine what the next offer should be and when to make it. This testing enables companies to carefully manage offers to avoid over-soliciting and possibly alienating their customers.

### Loyalty, Attrition, and Churn

A familiar mantra in marketing is, "It costs less to retain a customer than it does to replace one." Every company benefits from loyal customers. *Loyalty*, also known as *retention*, can be thought of as the opposite of *attrition*—the loss of a customer relationship. For highly competitive industries where markets are close to saturation, these terms are particularly relevant. Industries that count on renewals, such as insurance, telecommunications, and publishing, pay close attention to retention and attrition.

*Attrition* is defined as a decrease in the use of a product or service. For credit card banks, attrition is the decrease in balances on which interest is earned. It occurs when customers switch companies, usually to take advantage of a better deal. For years, credit card banks lured customers from their competitors by offering low interest rates. Telecommunications companies continue to use strategic marketing tactics to lure customers away from their competitors. Many other industries spend much effort to retain customers and steal new ones from their competitors.

For several decades, credit cards have been offered to almost every segment of the population, resulting in a saturated market. This reality means that credit card banks are now forced to increase their customer base primarily by luring customers from other providers. Their tactic is to offer low introductory interest rates for anywhere from three months to one year or more on either new purchases or balances transferred from another provider, or both. Their hope is that customers will keep their balances with the bank after the interest converts to the normal rate. Many customers, though, are savvy about keeping their interest rates low by moving balances from one card to another before the rate returns to normal. These activities introduce several modeling opportunities. One kind of model predicts the customer's reduction or ending of the use of a product or service after an account is activated.

In some industries, attrition becomes cyclical and is known as *churn*, which is defined as customers' closing of one account in conjunction with the opening of another account for the same product or service, usually at a reduced cost. Companies have learned that it is cheaper to keep a customer happy than to replace him or her. To this end, retention and attrition modeling have become popular in some industries.

Businesses use social network analysis to predict churn by examining the group behavior. By measuring the density of connection in a mobile phone network, for example, cellular companies are able to predict patterns of churn and take proactive steps to retain their customers.

## Risk and Approval Models

*Risk models,* or *approval models,* are unique to industries that assume the potential for loss when offering a product or service. The most familiar kinds of risk occur in the banking and insurance industries. However, there is a new focus on models to predict risk in the form of fraud and security breaches. As a result of the growth of data and interconnectivity, the level of risk related to data and technology is increasing exponentially.

## Default

*Default models* have been used to determine risk for several decades. On the basis of an individual's or business's credit history, financial profile, demographics, or similar data, a lending institution or other business can determine its applicant's likelihood of meeting his or her loan obligation. Predicting the likelihood of default or bankruptcy has been a main profit-driver of the loan industry for many years. Whether it's for a mortgage, credit card, car loan, or even utility service, many companies rely on a credit score to determine approval and terms of a contract. Banks also use aggregations of the expected loss to meet regulatory requirements for loan reserves.

## Loss-Given-Default

At the point of default, *loss-given-default models* are used to estimate the size of the loss. Some loans, such as mortgages and automobile loans, are secured, meaning that the bank holds the title to the home or automobile for collateral. The risk is then limited to the loan amount minus resale value of the home or car. Unsecured loans are loans for which the lender holds no secured assets, such as credit cards.

Many other industries incur risk by offering a product or service with the promise of future payment. This category includes telecommunications companies, energy providers, retailers, and many others. The type of risk is similar to that of the banking industry in that it estimates the probability of a customer's defaulting on the payment for goods or services.

## Claim

For the insurance industry, the risk is that of a customer's filing a claim. The basic concept of insurance is to pool risk. Insurance companies have decades of experience in managing risk. Insurance companies use predictive model scores to determine approval, optimize pricing, and determine reserve levels for life, health, auto, and homeowners insurance on the basis of demographics, risk factors including claim history, and credit risk factors. Because of heavy government regulation of pricing in the insurance industry, managing risk is critical to insurance companies' profitability. Warranty companies often use product information and safety records to model the likelihood of a claim.

## Fraud

With the increase in electronic and online transactions, fraud is increasing at an alarming rate. According to fraud research by the 2012 Global Fraud Study, survey participants estimated that the organizations lose an average of 5% of their revenues to fraud each year (Association of Certified Fraud Examiners 2012). Applied to the 2011 gross world product, this figure translates to a potential projected annual fraud loss of more than $3.5 trillion. Because of fraudulent purchases made with stolen credit cards, fake insurance claims (including life, health, and auto), stolen cellular usage, false tax returns, account fraud through Automated Clearing House, and money laundering, fraud increases costs for businesses and consumers alike.

Powerful models are in use and in development to thwart these costly actions.

Credit card banks use predictive models to identify the types of purchases that are typically made with a stolen card. For example, if the card is used to purchase expensive jewelry, furs, or firearms, the bank will freeze the card until the cardholder can verify the charges.

Insurance claim fraud is also costly to insurers and policyholders. Historically, every claim was personally inspected for fraud. Today, predictive models can estimate the likelihood that a claim is fraudulent based on common characteristics of past fraudulent claims. This model allows insurers to reduce expenses by automating the approval of low-risk claims and placing a priority on claims with a high likelihood of fraud.

Government institutions, including the Internal Revenue Service and social service agencies such as Medicare and the Social Security Administration, use predictive models to identify false tax returns or requests for benefits.

## Lifetime Value

Lifetime value (LTV) is the expected value of a prospect or customer over a specified period of time, measured in today's dollars. When calculated with the use of a series of models or other estimates, it allows a company to make decisions that optimize overall profits.

LTV is measured in various ways, depending on the industry, but basically it represents the value of future revenues minus overhead and expenses. It can be estimated from a series of estimates from models or other sources that may consider one or more of the following: response, activation, retention, risk, cross-sell, up-sell, and revenue. This evaluation enables companies to allocate resources based on customer value or potential customer value.

Historically, marketing strategies were driven by the financial benefits of a single campaign. Customer profitability was optimized by the net profits of the initial sale. With the increased cost of acquiring customers and the expansion of products and services to existing customers, companies are expanding their marketing strategies to consider the LTV of a potential customer.

LTV measurements on a customer portfolio can quantify the long-term financial health of a company or business. However, focusing on long-term goals can be challenging for companies if they suboptimize short-term profits.

The customer life cycle comprises three main phases (Figure 1.1):

- prospect
- new or established customer
- former or lapsed customer

Many opportunities for developing predictive models exist within the life cycle.

**Figure 1.1: Customer Life Cycle**

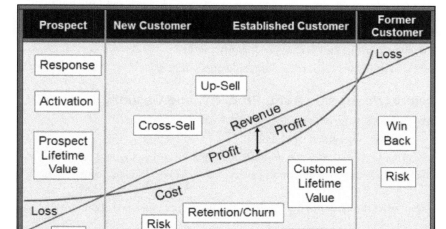

A prospect can be modeled for the propensity to respond or to activate. The risk level can also be estimated with predictive modeling or segmentation. By combining response, activation, or risk models with some customer value estimates such as retention (churn) or subsequent revenues, companies can calculate a prospect's LTV.

After a prospect becomes a customer, numerous additional modeling opportunities exist. Finally, after the customer relationship ends, models can be developed to improve profitability. By integrating models into a customer relationship management program, companies can reduce expenses while improving the efficiency of their marketing efforts.

## Predictive Modeling Opportunities by Industry

In the past few decades, predictive models have gained popularity across a variety of industries. Statisticians and data miners have found that the methods pioneered by the financial services industry have broad applicability to most other industries.

For all of the industries listed in this section, a combination of factors drives profitability. The financial impact of these factors can be predicted with the following types of models:

### Banks: Credit Card, Retail, Mortgage, and Savings and Loan

- *response model*—to target offers for loans
- *default model*—to predict non-payment or bankruptcy

- *revenue model*—to estimate the amount of purchases and earned interest
- *retention/loyalty/attrition model*—to estimate the optimal combination of pricing and services
- *cross-sell/up-sell model*—to increase usage on credit cards or purchase additional products and services such as investment products, retail banking services (savings and checking accounts), refinance of mortgage, etc.
- *fraud model*—to detect lost or stolen cards or identity theft
- *collection model*—to prioritize resources in the collection of debt

## Insurance Companies: Life, Health, Auto, Property, and Casualty

- *response model*—to target offers for insurance
- *default model*—to predict likelihood of claims
- *retention/loyalty model*—to estimate the optimal combination of pricing and services
- *cross-sell/up-sell model*—to increase insurance coverage (life) or purchase other types of insurance
- *fraud model*—to predict the likelihood of a false insurance application or claim

## Retailers: Catalog Companies and Retail

- *response model*—to target offers for goods and services
- *catalog companies and offline retailers*—to reduce snail mail costs
- *online retailers*—to improve targeting
- *revenue model*—to estimate the amount of the purchases
- *retention/loyalty model*—to estimate the optimal combination of marketing actions and incentives
- *cross-sell/up-sell model*—to increase repeat sales

## Telecommunications Companies: Cable, Internet, Wired-phone, and Cellular

- *response model*—to target offers for phone or internet service
- *revenue model*—to estimate the amount of usage
- *retention/loyalty model*—to estimate the likelihood of cancellation or churn
- *default model*—to predict the likelihood of defaulting on payment
- *cross-sell/up-sell model*—to increase usage and optimize product and service bundling
- *fraud model*—to predict illegal access to services
- *process failure model*—to predict grid or supply-chain failure

## Entertainment and Social Media Industry: Radio, Television, Social Media, and Online Sharing Sites

- *response model*—to increase click-through, views, and purchases based on variations in ad content, timing, and placement

- *revenue model*—to estimate the amount of the purchases
- *retention/loyalty model*—to estimate the optimal combination of marketing actions and incentives
- *failure model*—to predict internet or media delivery interruptions

## Gaming Industry: Casinos and Online Gaming

- *response model*—to optimize ad placement and incentives
- *retention/loyalty model*—to estimate the optimal combination of marketing actions and incentives; for example, many casinos have strong loyalty programs that include food, lodging, and priority access
- *cross-sell/up-sell model*—to increase usage and optimize product and service bundling
- *fraud model*—to predict the likelihood of a false representation, such as a stolen credit card or identify theft

## Education: Colleges, Universities, Technical Colleges, and Art Schools

- *response model*—to target an offer based on the likelihood of acceptance
- *retention model*—estimate the likelihood of graduation

## Technology Companies: Computer Hardware and Software

- *response model*—to target offers for product or service
- *revenue model*—to estimate the amount of usage for services
- *retention/loyalty model*—to estimate the likelihood of repeat purchases
- *cross-sell/up-sell model*—to increase usage and optimize product and service bundling
- *process failure model*—to predict product, operations, or supply-chain failure

## Utilities: Gas, Electric, and Water

- *response model*—to target offers for services in deregulated areas
- *usage model*—to estimate the amount of usage.
- *retention/loyalty model*—to estimate the likelihood of cancellation or churn
- *default model*—to predict the likelihood of defaulting on payment
- *cross-sell/up-sell model*—to increase usage and optimize product and service bundling

- *fraud model*—to predict illegal access to services
- *failure model*—to predict grid or supply-chain failure

## Hospitality and Travel Industry: Hotels, Resorts, and Restaurants

- *response model*—to target offers for phone or internet service
- *usage model*—to estimate the number of potential customers in order to plan staffing and supplies
- *revenue model*—to estimate the amount of revenue
- *retention/loyalty model*—to estimate the likelihood of repeat customers
- *cross-sell/up-sell model*—to increase usage and optimize product and service bundling
- *process failure model*—to predict operations or supply-chain failure

## Manufacturing: Auto, Pharmaceutical, Biotech, Chemical, Food, etc.

- *response model*—to predict sales
- *usage model*—to predict product expiration, disease outbreak, poisoning, allergic reaction
- *revenue model*—to predict production needs
- *process failure model*—to predict a breakdown in operations or a supply-chain
- *product failure model*—to predict the likelihood of a product malfunction or safety recall

## Public Sector and Nonprofit: Government, Political, Religious, Community, Education, Healthcare

- *response model*—to predict who will donate or help with campaign
- *revenue model*—to estimate the amount of donations
- *target model*—to predict for whom someone will vote
- *fraud models*—to detect false charges for services, identity theft, or IRS tax filings
- *usage models*—to plan staffing and supplies for hospitals, emergency centers, first responders, or nursing homes

## Transportation and Shipping: Rail, Ship, Truck, and Airline

- *usage model*—to predict quantities, sales, and fuel costs
- *geo usage model*—to optimize routes and storage locations
- *process failure model*—to predict operations or supply-chain failure

## Sports

- *recruitment model*—to predict whether a player's statistics are a good fit for the team
- *target model*—to predict who will attend an event

## Notes from the Field

The fact that you are reading this book suggests that you are among those of us who are seduced by the power and beauty of big data and its emergent insights. However, big data has a dark side. One can easily become lost in the minutia and forget about the main purpose of the project—that carefully crafted business goal.

In any data analysis or modeling project, you will benefit from stepping back and looking at your work from a leader's perspective. Typically, the CEO doesn't have the skills or the time to explore a project's details to the extent that you do. So, sometimes you must play CEO for your part of the business.

Engaging powerful tools like SAS Enterprise Guide and SAS Enterprise Miner is like sitting behind the wheel of a Ferrari. These tools can take you far and fast, but will you end up where you need to go?

To reach your destination in sound condition, you must know what the goal is, and you must have a map to direct your progress. You must also know where you are now and what your assets are: your data, tools, reference materials, and support resources.

If you are a manager, discipline and focus will be your best tools during this exciting process. When you are performing data analysis, allow your business acumen to guide you. The insights that emerge can serve to validate and enhance your knowledge, as well as inform your strategy.

If you are an analyst, you have a real opportunity here. The skills that brought you to this position are highly valued in the marketplace. But to become a true asset to your organization, you must learn the business—from strategy to operations. Your understanding of the business will help you make decisions based not only on the data, but also on what your findings mean to the business. If you're developing analysis and models to guide your strategy, then you will want to think through all the ramifications of your strategic decisions as they relate to the goals of the business.

# Chapter 2: Data Types, Categories, and Sources

## Introduction

Data is being created at an increasingly exponential rate. It's no wonder that, in today's high-tech, interconnected global economy, data is a foundational asset for all Fortune 500 organizations, as well as an increasing number of smaller companies.

In this chapter, you will learn about the different types and sources of data available for modeling and analysis. This data is the basis of your analysis. In the end, your final deliverable, anything from a simple report to an advanced model algorithm, is only as good as your data.

The chapter begins with a brief historical perspective on the evolution of big data collection and usage in business. This foundational knowledge not only gives you what you need to begin building a base for modeling and analysis, but will also teach patterns of data creation and usage that will enable you to be on the lookout for new types and sources of data.

# The Evolution of Data

Financial institutions were the early pioneers in the use of big data to manage risk and optimize marketing. Today, the use of data to drive decision-making has moved into all departments of most Fortune 500 businesses. Balanced Scorecard and Six Sigma are powerful tools for measuring and managing business processes. The next area of analysis with great potential for improving business is the measurement of human capital and performance.

To leverage this data explosion, companies are diligently collecting, cleaning, combining, storing, and analyzing their data. Many companies are making major investments in enterprise data management systems to ensure accuracy and accessibility. To understand the real significance of this increased focus on data, you will find that looking at some recent history is worthwhile.

Since the early 1990s, advances in computer power, server capacity, and connectivity have fueled the growth of data collection and usage. This growth has enabled companies to market directly to consumers and businesses. Companies have emerged with the sole purpose of collecting data on individuals and businesses from a variety of public and private sources, such as U.S. Census statistics, phone directories, warranty cards, and public records. A long history of credit behavior also has been available for credit-related use.

As companies began to realize the financial benefits of utilizing these sources, new businesses formed for the sole purpose of collecting, cleansing, aggregating, enhancing, and selling data on individuals and businesses. Today, the dramatic increase in World Wide Web traffic, social media, and mobile devices has generated enormous amounts of data. And each new technology, process, or activity is a potential source of new data. Experts are calling this phenomenon *big data*.

Large data sets have been around for many years. But the growth in the amount of data is now increasing exponentially. Many factors contribute to the increase in the amount of data being generated: transaction-based data stored through the years, unstructured data streaming in from social media, and increasing amounts of sensor and machine-to-machine data being collected. In the past, excessive data volume was a storage issue. But with decreasing storage costs, other issues emerge, including how to determine relevance within large data volumes and how to use analytics to create value from relevant data.

But big data is also characterized by how quickly it changes and its numerous sources. Data today comes in many different formats: structured, numeric data in traditional databases; information created from line-of-business applications; and unstructured text documents, email, video, audio, stock ticker data, and financial transactions. Managing, merging and governing varieties of data is still a struggle for many organizations.

According to SAS Institute in "Big Data: What It Is and Why It Matters," data earns its "bigness" in two additional ways (SAS Institute Inc., 2014):

- **Variability.** In addition to the increasing velocities and varieties of data, data flows can be highly inconsistent, with periodic peaks. Is something trending in social media? Daily, seasonal, and

- event-triggered peak data loads can be challenging to manage, and even more so when unstructured data is involved.
- **Complexity.** Today's data comes from multiple sources, and it is still an undertaking to link, match, cleanse, and transform data across systems. However, it is necessary to connect and correlate relationships, hierarchies, and multiple data linkages, or your data can quickly spiral out of control.

New kinds of data from digitization and social media are adding to the "bigness" of data. Unstructured data appears to be the next frontier of data sources. Powerful software is now able to decipher and analyze handwritten text, audio, images, and videos. The translation of unstructured data and its integration with structured data are projected to be the primary sources of data for analysis in the future.

The first step in making the best use of any data source is to understand the nature of the data, how it is gathered, and how it is managed.

## Types of Data

Data comes in three basic types: nominal, ordinal, and continuous.

### Nominal Data

*Nominal data* is qualitative or descriptive data that has discrete values.

Variable or data columns that contain nominal data are often referred to as *class variables*, *categorical variables*, or *discrete variables*. Nominal data can be used for segmentation or classification.

When a class variable is used as an input to a particular kind of predictive model, such as a regression or neural network model, it must be converted into a numeric form. For example, you could say that the values for gender are 1 and 0, where 1 = *male* and 0 = *female*. When the values are set to 0 and 1, they are also referred to as *indicator variables*. This transformation can be accomplished automatically in modeling software such as SAS Enterprise Miner.

### Ordinal Data

*Ordinal data* is data with categories that *have* relative importance. The categories can be used to rank strength or severity. For example, you may be familiar with the rankings "Low, Medium, and High." These have a relative ranking but can't be compared mathematically.

Numeric values can also be used to denote relative ranking without a mathematic relationship, For example, a list company assigns the values 1 through 5 to denote financial risk. The value 1, characterized by no late payments, is considered low risk. The value 5, characterized by a bankruptcy, is considered high risk. The values 2 through 4 are characterized by various previous delinquencies. A prospect with a risk ranking of 5 is definitely riskier than a prospect with a ranking of 1, but he or she is not five times as risky. And the difference in their ranks, $5 - 1 = 4$, has no meaning.

## Continuous Data

*Continuous data,* or *ratio data,* is the most common data used to develop predictive models. It can accommodate all basic arithmetic operations: addition, subtraction, multiplication, and division. Most business data, such as sales, balances, and minutes, is ratio or continuous data.

## Categories of Data

Three main characteristics of data should be considered when you are deciding the types of data you want to use: the power, stability, and cost. First, however, you will want to look at the data categories.

Although data is available from numerous sources and is of varying types, regardless of its origin, data fits into one of three basic categories: demographic, behavioral, or psychographic or attitudinal. Each category has its strengths and weaknesses.

### Demographic or Firmographic Data

*Demographic data* generally describes personal or household characteristics. It typically includes the following characteristics:

- gender
- age
- marital status
- income
- home ownership
- dwelling type
- education level
- ethnicity
- presence of children

Demographic data has many strengths. It is stable, which makes it appealing for use in predictive modeling. Characteristics like marital status, home ownership, education level, and dwelling type aren't subject to change as frequently as behavioral data, such as bank balances, or attitudinal characteristics, like favorite political candidate. And demographic data is usually less expensive than attitudinal and behavioral data, especially when purchased on a group level.

One of the weaknesses of demographic data is that it is difficult to collect on an individual basis with a high degree of accuracy. Unless it is required by law or is required in return for a product or service, many people resist sharing this type of information, or they supply false information.

*Firmographic data* can be thought of as demographic data for businesses. It includes characteristics, such as industry, annual revenue, number of employees, number of sites, and growth rate. It has many of the same features as demographic data. However, it is often more accurate on an individual level, especially for publicly traded companies.

## Behavioral Data

*Behavioral data* is a measurement of people's action or lack of action. Typically, behavioral data is the most predictive kind. Depending on the industry, behavioral data may include elements such as the following:

- response to an offer
- purchase date
- amount of purchase
- type of purchase
- ending of subscription
- payment date
- payment amount
- customer service action
- insurance claim
- default
- bankruptcy

Website and social media activity are also forms of behavioral data. Websites can track the path of a visitor. Social media sites are able to collect an enormous amount of information based on your preferences and connections.

Social network data is a type of behavioral data that includes size and density of a social group, its stability, and within-group similarity. Members of a social network are evaluated on characteristics such as connectedness, social role, and traffic role.

You can collect online behavioral data by analyzing the sequence and timing of visitors to a website or series of websites. The results can be used to assist with search engine optimization and increase website "stickiness" or the time a visitor lingers on a page or site.

Behavioral data is better for predicting future behavior than the other categories of data. It is, however, generally the most difficult and expensive data to get from an outside source. Consequently, most companies focus on internally tracking and storing their customer data.

**TRENDS:** Web-savvy businesses are developing predictive models by using online patterns of behavior to estimate preferences, industry changes, untapped markets, and future trends. Using estimates from these models, business can measure sentiment and activities that can inform timing and messaging of marketing or other actions.

## Psychographic Data

*Psychographic data* is also known as *attitudinal data* or *lifestyle data*. It is characterized by opinions, fashion, style, political choices, personal values, or some combination. Traditionally associated with market research, this category of data is collected mainly through surveys, opinion polls, and focus groups. It can also be inferred through subscription, purchase, and social media behavior. Because of increased competition, this category of data is being integrated into customer and prospect databases for improved target modeling and analysis.

Psychographic data brings an added dimension to predictive modeling. For companies that have squeezed all the predictive power possible out of their demographic and behavioral data, psychographic data can offer some improvement. It is also useful for determining the life stage of a customer or prospect. This data creates many opportunities for developing products and services around life events such as marriage, childbirth, college expenses, and retirement.

The biggest drawback to psychographic data is that it denotes intended behavior that may be highly, partly, or marginally correlated with actual behavior. Data may be collected through surveys or focus groups and then applied to a larger group of names, such as a customer base, using segmentation or predictive modeling. If data is applied with the use of these methods, it is recommended that a test be constructed to validate the correlation.

## Data Category Comparison

Table 2.1 provides a quick reference and comparison of the three main categories of data. The rating is based on individual-level data. If data is collected on a group and inferred on an individual level, it is generally less predictive and less expensive. The stability remains the same.

**Table 2.1: Data Category Characteristics**

| Data Category | Power | Stability | Cost |
|---|---|---|---|
| Demographic | Medium | High | Low |
| Behavioral | High | Low | High |
| Psychographic | Medium | Medium | Low or High |

The cost for psychographic data shows *Low or High* because social media data is becoming a low-cost source of this data. As tools become available to access and mine the rich information in social media data, the cost drops accordingly.

# Sources of Data

Data for modeling can be generated from numerous sources. These sources fall into one of two categories: internal or external.

Internal sources are those that are generated through company activity and are stored within the company, typically in a data warehouse. They include, but are not limited to, customer records, transactional data, World Wide Web behavior, and marketing data (both offer and customer behavior). External sources of data include companies such as credit bureaus, list brokers and compilers, and corporations with large customer databases, such as retailers, publishers, and social media sites.

## Internal Sources

Internal sources are optimal for analysis because they represent information that is specific to the company's product or service. The following subsections detail the typical features and components of these databases.

### Customer Database—Business to Consumer

A business to consumer (B2C) customer database, designed to target consumers, typically is designed with one record per customer. In most organizations, it is one of several databases. It contains identifying information that is linked to other databases, such as a transaction database to obtain a current "snapshot" of a customer's performance. Even though wide variation among companies and industries may exist, following are some key elements in a typical customer database:

- *Customer ID* is a unique numeric or alphanumeric code that identifies the customer throughout his or her entire customer life cycle. Some companies may use an account number for this function, but doing so can be risky if the account numbers are subject to change. For example, credit card banks assign a new account number when a card is lost or stolen. The customer ID allows each account number to be linked to the unique customer, thereby preserving the entire customer history. It is essential in any database to effectively link and track the behavior of and actions taken on an individual customer.

- *Household ID* is a unique numeric or alphanumeric code that identifies the household of the customer through his or her entire customer life cycle. This identifier is useful in some industries in which more than one member of a household shares products or services.

- *Account number* is a unique numeric or alphanumeric code that relates to a particular product or service. One customer can have several account numbers.

- *Customer name* is the name of a person or a business. It is usually broken down into multiple fields: last name, first name, middle name or initial, salutation.

- *Physical address* is typically broken into components such as number, street, suite or apartment number, city, state, postal code, and country. Some customer databases have a line for a P.O. Box. With population mobility about 10% per year, additional fields that contain former addresses are useful for tracking and matching customers to other files.

- *Phone number* includes area code, country code, and, if applicable, extension.

- *Email address* can also serve as unique identifier.

- *Demographic information* such as gender, age, and income may be stored for profiling and modeling.

- *Products or services* are typically tracked with the use of a list of products and product identification numbers that may vary by company. An insurance company may list all its insurance policies, together with policy numbers. A bank may list all the products across different divisions of the bank, including checking, savings, credit cards, investments, loans, and more. If the number of products and the product detail are extensive, this information may be stored in a separate database with a customer and household identifier.

- *Model scores* such as response, risk, retention, profitability scores, or any other scores may be purchased or developed in-house. These scores may reside in a separate database that is linked by the account number.

## Customer Database—Business to Business

A customer database designed to target businesses (B2B) resembles a customer database designed to target consumers (B2C). The main difference is that the customer is a business instead of an individual.

The overall structure is also similar. The main database typically has one record for each company unit. Each record or site may also be a headquarters (HQ) or a map to another site that is the HQ. Each record contains the identifying information that can be linked to other databases, such as a transaction database, a marketing database, or other databases that track past performance and other information of value to the company.

In addition to a company-level database, many large B2B companies have a contact database in which the individuals are associated with companies in the customer database. It may also contain additional contacts that work for companies that are not yet customers.

> **NOTE:** Some companies sell to both consumers and businesses. Their databases are either separate or are a hybrid of both business and consumer data.

Even though there may be wide variation among companies and industries, the following are key elements in a typical customer database:

- *Company ID* is a unique numeric or alphanumeric code that identifies the company throughout its life cycle.
- *Headquarter ID* is a unique numeric or alphanumeric code that identifies the headquarter (HQ) of the company throughout its entire life cycle.
- *DUNS number (site)* is a unique ID that is available through Dun and Bradstreet (D&B). This ID allows for mapping to firmographic data provided by D&B.
- *DUNS number (other)* is one of a series of numbers that can match each site to its HQ, domestic ultimate (DU), or global ultimate (GU). It may be the same as the site ID if the site is also the HQ, DU, or GU.
- *Company name* is the name of the business. For a sole proprietor, it may be a person's name.
- *Physical address* is typically broken into components such as number, street, suite or apartment number, city, state, postal code, and country. Business addresses are more stable than addresses for individuals, especially companies with many employees.
- *Phone number* is the main number for the organization and typically includes area code and country code. If the data is at a contact level, the phone number may include area code, country code, and, if applicable, extension.
- *Company URL or web address* can also serve as a unique identifier.
- *Firmographic data* resembles demographic data but is at the company level. It includes industry, number of employees, age of company, annual sales, growth rate, standard industrial classification codes, and the like.
- *Product or service IDs* typically consist of a list of identification numbers that vary by company. Product or service IDs may reside in a separate database that is linked by the company ID.
- *Model scores* such as response, risk, retention, profitability scores, or any other scores may be purchased or developed in-house. These scores may reside in a separate database that is linked by the company ID.

## Contact Database

The contact database contains data about the employees of companies found in the company database, as follows:

- *Contact ID* is a unique numeric or alphanumeric code that identifies the contact throughout his or her entire life cycle.
- *Contact name* is the name of a customer or prospect. It is usually broken down into multiple fields: last name, first name, middle name or initial, and salutation.

- *Company ID* is a unique numeric or alphanumeric code that identifies the company throughout its life cycle.
- *Company name* is the name of the business. For a sole proprietor, it may be a person's name.
- *Physical address* is the street address, typically broken into components such as number, street, suite or apartment number, city, state, postal code, and country. Companies with virtual workers can have contacts anywhere in the world.
- *Phone number* includes area code, country code, and, if applicable, extension.
- *Email address* can also serve as unique identifier. *Job title or job level* describes the role, department, and level of each contact.
- *Model scores* such as response, risk, retention, profitability scores, and any other scores may be purchased or developed in-house. These scores may reside in a separate database that is linked by the contact ID.

## Transaction Database

For certain industries, the transaction database contains records of customer activity. It is often the richest and most predictive information, but it can be the most difficult to use. In most cases, each record represents a single transaction, so multiple records may exist for each customer.

The transaction database can take on various forms, depending on the type of business. For you to use this data for modeling, it must be summarized and aggregated to a customer level. Number of records per customer can differ. The following typify what might be found in a transaction database:

- *Customer ID or contact ID* is a unique numeric or alphanumeric code that identifies the customer throughout his or her entire life cycle. Some companies may use an account number for this function.
- *Account number* is a unique numeric or alphanumeric code that relates to a particular product or service.
- *Activity type* provides detail about the type of activity, such as purchase, payment, fee, or return.
- *Activity amount* is the amount of the transaction or activity.
- *Activity date* is the date that the transaction occurred.

These fields must be summarized to the customer or company level for most analyses.

## Score Database

Many companies develop models for all areas of their business. The scores from these models can be stored in a score database. Some large companies may have multiple databases that contain different types of scores. For example, a bank might have one score database for risk scores, such as the probability of default or the potential loss upon default. That same bank might have a second score database for marketing scores, such as the probability to take a loan, carry high balances, or take an additional product or service.

The following typify what might be found in one or more score databases:

- *Customer ID* or *contact ID* is a unique numeric or alphanumeric code that identifies the customer throughout his or her entire life cycle. Some companies may use an account number for this function.
- *Probability to buy* is a score that represents the likelihood of a customer to purchase a particular product or service. This score can be for a new purchase or an additional purchase from the same company. It may cover a range of products and services.
- *Expected value* is the amount, in dollars or numbers, that the customer or company is expected to purchase or borrow, if response occurs.
- *Probability to default* is the likelihood that a customer will not pay the amount owed.
- *Expected loss given default* is the amount that a customer or company is expected to have as a balance when default on the loan occurs.

## Marketing Database

The marketing database contains details about offers made to both prospects and customers. The most useful format is a unique record for each campaign to each customer or prospect.

A marketing database would contain all cross-sell, up-sell, and retention offers. A prospect marketing database would contain all acquisition offers, as well as any predictive information from outside sources.

Variables created from this type of database are often the most predictive for targeting models. It seems logical that if you know that someone has received your offer every month for six months, then he or she is less likely to respond than someone who is seeing your offer for the first time. As competition intensifies, this kind of information is becoming increasingly important.

With an average amount of solicitation, this kind of database can become huge. Consequently, you should perform analyses to establish business rules that control the maintenance of this database. Fields like *date of first offer* are usually correlated with response behavior. Following are key elements in a marketing database:

- *Prospect ID or customer ID* is a unique numeric or alphanumeric code that identifies the prospect for a specific length of time.
- *Offer detail* includes the date, type of offer, marketing copy, source code, pricing, distribution channel (mail, telemarketing, sales representative, email), and any other details of the offer.
- *Date* can be date of first offer (for each offer type) or other offer dates (unique to product or service).

## Storage of Data

### Data Warehouse

A *data warehouse* is a structure that links information from two or more databases. As shown in Figure 2.1, the data warehouse brings the data into a central repository, performs some data integration, cleanup, and summarization, and distributes the information into data marts. Data marts are used to house subsets of the data from the central repository. The data in data marts is selected and prepared for specific end users.

**Figure 2.1: Data Warehouse**

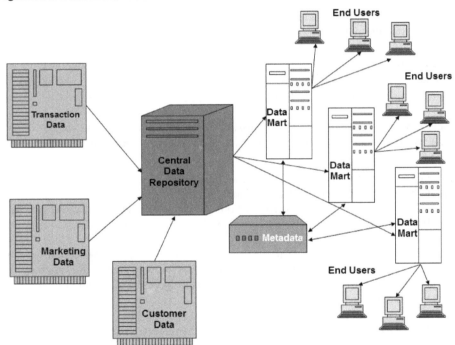

### Hadoop

Apache™ Hadoop is revolutionizing the leveraging of big data. Hadoop is a data platform that operates by dispersing data storage and processing over a large group of servers. Hadoop can quickly scale data processing from a single machine to several thousand machines.

Traditional models of data storage are expensive, especially for large data sets. Hadoop is built at a much lower cost and uses commodity hardware. By operating at the petabyte scale, Hadoop can reduce costs tremendously. This savings will continue to increase as the cost of hardware continues to decrease.

Companies use web data to understand customer behavior and improve marketing. But the immense volume of web data can be prohibitive to capture and store. Hadoop allows companies to capture and store web data at a much lower costs. It also allows companies to hold onto historical data for a longer period of time.

Hadoop also offers the ability to analyze large volumes of data. Because of the distributed servers, Hadoop can process large volumes of data in parallel.

The ability to perform advanced analytics is a key feature of Hadoop. Hadoop also has the capacity to support several higher-level business intelligence tools.

## External Sources

The pressure is on many companies to increase profits either by acquiring new customers, or by increasing sales to existing customers. You can enhance both of these initiatives with the use of external sources.

External sources consist mainly of list sellers and compilers. Few companies that sell lists, however, have these sales as their sole business. Many companies have a main business, like magazine sales or catalog sales, with list sales as a secondary business. Depending on the type of business, they usually collect and sell names, addresses, email messages, and phone numbers, together with demographic, behavioral, and psychographic information. Sometimes they perform list "hygiene," or cleanup, to improve the value of the list. Many of them sell their lists through list compilers or list brokers.

List compilers are companies that sell a variety of single and compiled lists. Some companies begin with a base like the phone book or driver's license registration data. Then they purchase lists, merge them, and impute missing values. Many list compliers use survey research to enhance and validate their lists.

Many companies sell lists of names with contact information and personal characteristics. Some specialize in specific kinds of data. The credit bureaus are well known for selling credit behavior data. They serve financial institutions by gathering and sharing credit behavior data among their members. Hundreds of companies sell these lists, ranging from very specific to nationwide coverage.

**TIP:** Commonly, prospect data is purchased repeatedly from the same source. The goal may be to purchase new names or to get fresh information about existing names. In either case, you should arrange with the data seller to assign a unique identifier to each prospect that is housed on the seller's database. Then, when additional names are purchased or current names are refreshed, the match rate will be much higher than it otherwise would be.

**NOTE:** For privacy reasons, social media data is mainly collected and used with each social media site. Some social media activities can be purchased.

## Notes from the Field

In this chapter, you've seen how data for analysis comes in many forms and from multiple sources. Your first step in preparing the data for analysis is to understand how and from where all the data sets were extracted. When data is purchased or extracted from unfamiliar sources, you should request data dictionaries that contain a definition for every field. Be sure you know what each field means and how it is derived (if applicable). In addition, request counts, ranges, and simple statistics for each field, such as mean, median, minimum, maximum, and standard deviation.

Make friends with the people in your information technology department. They can become valuable collaborators. They can help you get the names of the people in charge of creating the data, if possible. Don't hesitate to ask questions until you are sure you understand the data.

Data can be seductive. If you are starting a new project, you may be eager to collect as much data from as many sources as possible. However, with the amount of data available for analysis, too much of a good thing can be paralyzing.

The best approach is to work in stages. Begin with the question from Chapter 1: *What is your objective?* If you have many objectives, choose the most important two or three. Then, work with your data management team to optimize the process. If operational or pricing considerations justify the collection a large amount of data, then focus on a few that will have the most impact. Remember that analysis is usually done in isolation, but the planning should be done in concert with the major stake-holders and end-users to ensure the optimal use of talent, time, and money.

Remember that, in the end, your analysis is only as good at your data. This fact is true for both the quality of the data and how well it aligns with the objective of your analysis.

# Chapter 3: Overview of Descriptive and Predictive Analyses

## Introduction

In this chapter, you will learn about different types of descriptive and predictive analyses that are commonly used in business today. Once your source data sets are loaded and you have all the supporting documents necessary to access and understand the data, you are ready to start your analysis.

Descriptive analysis consists of tables, charts, and graphs that unveil data patterns, trends, and relationships that can inform your strategies, decisions, and actions. Predictive analysis tools like regression, decision trees, and neural networks give you the power to prioritize and optimize a broad range of marketing, risk, and process decisions.

You will see examples of each type of analysis with guidelines to determine when to use descriptive analysis and when to use predictive analysis. The goal of this chapter is to help you see opportunities to increase your competitive advantage by interpreting both descriptive and predictive analyses and taking appropriate business actions.

# Descriptive Analyses

The purpose of descriptive analysis is to reveal the characteristics of your data. It can describe a single point in time or cover a trend over multiple time intervals. When first receiving data, you should start with a simple descriptive analysis such as a frequency for categorical variables, and a distribution analysis for continuous variables. Both sets of results can be displayed with tables and graphs.

## Frequency Distributions

A *frequency distribution* is a representation of the basic shape or distribution of a variable in a data set. The frequency distribution can be displayed in a table or a graph. The components of the table are different depending on whether you are viewing a class variable or a continuous variable.

The examples in this section are calculated on data taken from automobile research.

### Class Variables

A frequency distribution table for a class variable typically displays counts and percentages. Table 3.1 shows a simple frequency analysis in a table that displays the frequency counts and percentages of the variable *Type (of car)*. You can easily see that 262 cars, or 61.2%, of the cars represented by this data are *sedans*.

**Table 3.1: Frequency Distribution Table for the Class Variable Type (of Car)**

| Value | Frequency Count | Total Frequency (%) |
|---|---|---|
| Sedan | 262 | 61.2150 |
| SUV | 60 | 14.0187 |
| Sports | 49 | 11.4486 |
| Wagon | 30 | 7.0093 |
| Truck | 24 | 5.6075 |
| Hybrid | 3 | 0.7009 |

Figure 3.1 shows the same information in a bar chart. Although you can't see exact percentages in the bar chart, this view can be easier for a quick comparison. You may want to use both because some people prefer tables, while others prefer visual displays of data such as charts.

**Figure 3.1: Frequency Distribution Bar Chart for the Class Variable Type (of Car)**

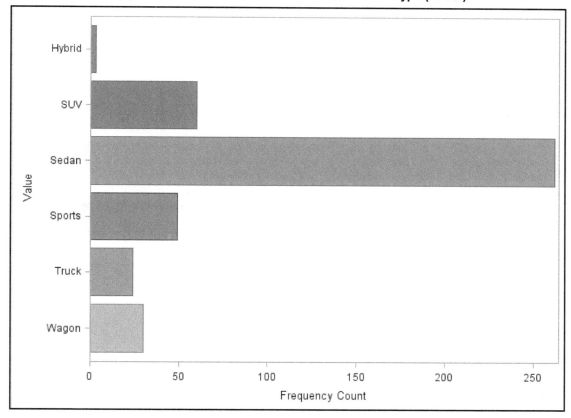

## Continuous Variables

A frequency distribution table for a continuous variable typically displays population measures, such as the values for mean, median, minimum, and maximum. Table 3.2 displays the values for the variable *miles per gallon (MPG) on the highway.* You can read the number of total observations in the data (428). No values are missing. The total of 11,489 is the sum of MPG for the whole set of cars represented here. For the variable *MPG_Highway*, the total may not be meaningful. On the other hand, if it were a variable representing bank account balances, the sum would represent the total balances for the population. This value could be of interest to a financial institution. Additional values represented in the table are the mean MPG with a value of 26.84. The minimum is 0, and the maximum is 66.

**Table 3.2: Frequency Distribution for the Continuous Variable MPG (Highway)**

| Category | Measure |
| --- | --- |
| Number (Nonmissing) | 428.00 |
| Number Missing | 11.00 |
| Total Number | 489.00 |
| Minimum | 12.00 |
| Mean | 26.84 |
| Median | 26.00 |
| Maximum | 66.00 |
| Standard Mean | 0.28 |

Figure 3.2 displays the frequency distribution in a bar chart. The values are bucketed to show the underlying nature of the data. When you view the frequency distribution bar chart of a continuous variable, you learn additional information, such as the skewness of the distribution.

**Figure 3.2: Frequency Distribution Bar Chart for the Continuous Variable Miles per Gallon (Highway)**

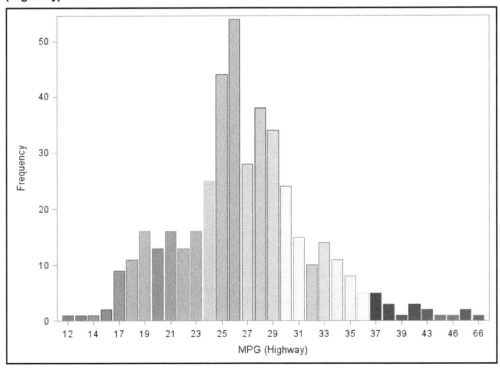

## Cluster

*Cluster analysis*, or *clustering*, is a data-driven process that groups observations into clusters that favor similarity within each cluster while favoring dissimilarity between clusters. The precise method used to calculate the similarities and differences will vary, depending on the goal of your analysis.

Figure 3.3 shows clusters that are not mutually exclusive. For example, one observation can populate more than one cluster. This cluster analysis shows segmentation by age and income. Notice that three groups form:

- low income and low age
- high income and medium age
- medium income and high age

**Figure 3.3: Example of Cluster Analysis with Clusters That Are Not Mutually Exclusive**

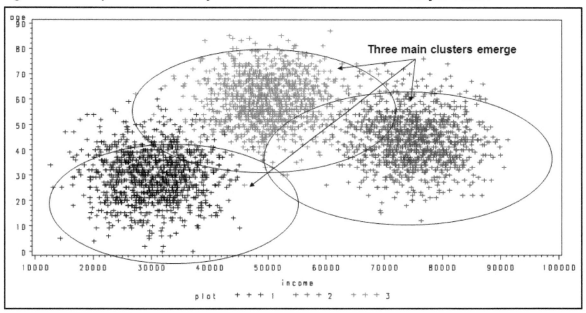

For use in marketing, the usual goal of cluster analysis is to separate people or companies into mutually exclusive groups. With SAS Enterprise Miner, clusters can be built into mutually exclusive segments, which are useful when segmenting a group of customers of companies. Chapter 7 includes an example of clustering into mutually exclusive segments.

## Decision Tree

Generating a decision tree is another method of understanding data relationships that can be used for both descriptive analytics and predictive analytics. Tree analysis sequentially partitions the data to maximize the differences in the dependent variable on the basis of the independent variables. It is also known as a *classification tree*. The true purpose of the tree is to *classify* the data into distinct groups, or branches, that create the strongest differentiation in the values of the dependent variable.

Decision trees are good at identifying segments with a desired behavior such as response or activation. This identification can be quite useful when a company is trying to understand what is driving market behavior. It also has an advantage over regression in its ability to detect nonlinear relationships that can be useful for identifying interactions for inputs into other modeling techniques.

A decision tree is "grown" through a series of steps and rules that offer great flexibility. In Figure 3.4, the tree differentiates between responders and nonresponders. The top node provides details of the size and the performance of an overall marketing campaign in which offers were mailed to 10,000 potential customers and yielded a response rate of 2.6%. When examining the results, you can see that the first split is on *gender*. This split implies that the greatest difference between responders and nonresponders is their gender. Males are much more responsive (3.2%) than females (2.1%). So you can describe the responders as more heavily male. From a predictive standpoint, after one split you would consider males the better target group. For additional descriptive and predictive insights, you can split the tree within each gender to find additional subgroups that further discriminate between responders and nonresponders. In the next split, the two gender groups or nodes are considered separately.

**Figure 3.4: Example of a Decision Tree for Target Marketing**

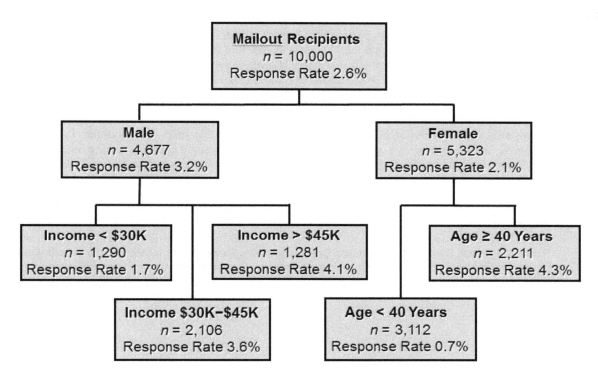

The second-level split from the *male* node is on *income*. This implies that income level varies the most between responders and nonresponders among the male customers. For female customers, the greatest difference is among *age* groups. So you can describe males with high incomes as among the best responders. You can describe females over age 40 as among the best responders. For predictive use, say that you decide to mail only to groups for which the response rate has been more than 3.5%; then, the offers would be directed to males who make more than $30,000 a year and to females older than age 40.

A definite advantage of decision trees over other techniques is their ability to explain the results. So if you develop a complex logistic model for scoring but it seems very hard to explain, a tree model might be helpful. Use the same variables from the complex model to build a tree model. Although the outcome is never identical, the tree does uncover key drivers in the market. Because of their broad applicability, decision trees will continue to be a valuable tool for all kinds of target modeling.

## Predictive Analyses

Today, numerous tools are available for developing predictive models. Some use statistical methods such as linear regression and logistic regression. Others use nonstatistical or blended methods, such as neural networks, classification trees, and regression trees. Much debate rages about which is the best method.

The steps surrounding the model processing are more critical to the overall success of the project than the technique used to build the model. This is the reason that the focus here is on regression. Logistic regression is available in any statistical software package, and it appears to perform as well as other methods, especially when used over a period of months or years.

## Linear Regression

In business, linear regression is useful for numerous measures that are in units of dollars or time. *Simple linear regression analysis* is a statistical technique that quantifies the relationship between two continuous variables: the dependent variable (the variable you are trying to predict) and the independent, or predictive, variable.

For example, Jake's Web Design tracked its advertising expenses and resulting sales for 15 months. The values are shown in Table 3.3.

**Table 3.3: Small Business Advertising Costs and Sales Revenue for 15 Months, in U.S. Dollars**

| Month | Cost of Advertising | Sales Revenue |
| --- | --- | --- |
| January | 120 | 1,503 |
| February | 160 | 1,755 |
| March | 205 | 2,971 |
| April | 210 | 1,682 |
| May | 225 | 3,497 |
| June | 230 | 1,998 |
| July | 290 | 4,528 |
| August | 315 | 2,937 |
| September | 375 | 3,622 |
| October | 390 | 4,402 |
| November | 440 | 3,844 |
| December | 475 | 4,470 |

| Month | Cost of Advertising | Sales Revenue |
|---|---|---|
| January | 490 | 5,492 |
| February | 550 | 4,398 |
| March | 120 | 1,503 |

Figure 3.5 is a graphical representation of the 14 months of data from Table 3.1. It represents sales figures, together with the U.S. dollar amount that was spent on advertising for Jake's Web Design. You can see a definite relationship between the two variables, *sales* and *advertising expenditure*; as the advertising expenditure increased, the sales also increased.

**Figure 3.5: Scatterplot Example of a Simple Linear Regression of a Small Business's Sales Revenue on Advertising Expenditures, in U.S. Dollars**

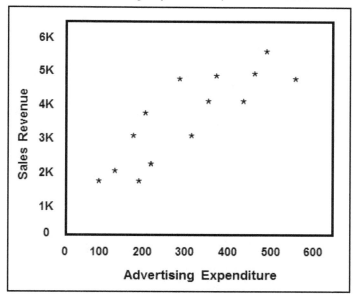

The goal of the model is to predict sales revenue on the basis of advertising expenditures. This technique works by finding the line that, when drawn through the data, minimizes the squared error from each point (Figure 3.6).

**Figure 3.6: Scatterplot Example, with Regression Line and Error Distance, of a Simple Linear Regression of a Small Business's Sales Revenue on Advertising Expenditures, in U.S. Dollars**

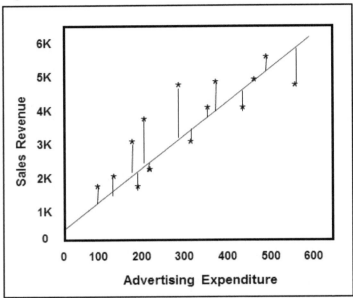

A regression analysis results in an equation that defines the relationship between the independent variables and the dependent variable. In our example shown in Figure 3.6, the equation *sales* = 866.59 + 7.81 × *advertising* indicates that the amount of expected sales is $866.59 plus 7.81 times the amount spent on advertising. A key measure of the strength of the relationship is the $R^2$. It measures the amount of the overall variation in the data that is explained by the model. This particular regression analysis results in an $R^2$ of 70%, meaning that 70% of the variation in *Sales Revenue* is explained by the variation in *Advertising Expenditure* in the regression model.

*Multiple linear regression* uses two or more independent, or predictive, variables to predict a continuous outcome variable. In Figure 3.7, *credit card balances* are a function of *payment amount* and *customer age.*

**Figure 3.7: Example of Multiple Linear Regression of Credit Card Balance on Payment Amount and Customer Age**

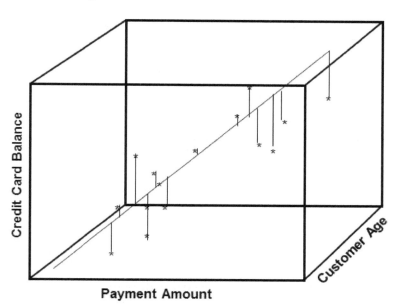

The equation *credit card balances* = 2.1774 + 9.4966 × *payment amount* + 1.2494 × *customer age* illustrates the predicted amount of credit card balances based on the customer's age and regular payment amount.

## Logistic Regression

*Logistic regression* resembles linear regression. The key difference is that the dependent variable is not continuous; it is discrete, or categorical. The discrete nature of the dependent variable makes it particularly useful in marketing because marketers often try to predict a discrete action, such as a response to an offer or a default on a loan.

Technically, logistic regression can be used to predict outcomes for two or more levels. When you build targeting models for marketing, however, the outcome is often bi-level. In these models, you predict the probability of one of these levels, and that probability must be from 0 to 1. A linear function of the predictor variables, however, is anywhere from negative infinity to positive infinity. Because logistic regression is a linear function of the predictor variables, a logit transformation is used to account for the difference between the two scales.

The goal in this section is to avoid heavy statistical jargon, but because logistic regression is a popular and powerful method for predictive analysis, an overview of the method is included. Remember that logistic regression resembles linear regression in the actual model processing.

Figure 3.8 shows a relationship between *response* (0/1) and *prospect's age*. The goal is to predict the probability of response to a catalog that sells high-end gifts using the *prospect's age*. Notice how the data points have a value of 0 or 1 for response. On the *prospect's age* axis, the values of 0 for *response* are clustered around the lower values for *prospect's age*. Conversely, the values of 1 for *response* are clustered around the higher values for *prospect's age*. A sigmoidal function, or *s* curve, is formed by averaging the 0s and 1s for each value of *prospect's age*. You can see that older prospects respond at a higher rate than younger prospects.

**Figure 3.8: Example of Simple Logistic Regression of Response (1 = Yes, 0 = No) to Sales Catalog on Prospect's Age (Years)**

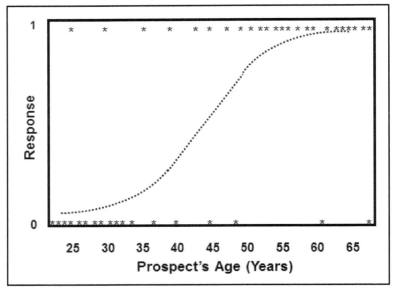

The following steps detail the construction of the logit function in simple language. This logistic regression example does not duplicate the data represented in the above graph, but it conveys the general idea:

1.  For each value of *prospect's age*, calculate a probability ($p$) by averaging the values of response.
2.  For each value of *prospect's age*, calculate the odds by using the formula $p/(1 - p)$.
3.  The final transformation calculates the log of the odds: $\log (p/(1-p))$.
4.  The parameter estimates are derived using maximum likelihood estimation, which produces the parameter estimates that are most likely to occur given the data. The parameter estimates are related to the log of the odds as follows: $\log (p/(1 - p)) = \beta_0 + \beta_1 X_1 + \beta_2 X_2 + \dots + \beta_n X_n$, where $\beta_0 \dots \beta_n$ are the parameter estimates and $X_1 \dots X_n$ are the predictive variables.
5.  After the parameter estimates, also called coefficients or weights ($\beta$s), are derived, the final probability is calculated using the following formula: $P = \exp(\beta_0 + \beta_1 X_1 + \beta_2 X_2 + \dots + \beta_n X_n)/(1 +$

exp ($\beta_0 + \beta_1 X_1 + \beta_2 X_2 + \ldots + \beta_n X_n$)). This formula is also written in a simpler form as follows: $p = 1/(1 + \exp(-(\beta_0 + \beta_1 X_1 + \beta_2 X_2 + \ldots + \beta_n X_n)))$.

6. The process considers the average value for response at each value of age. Values for $p$ are bounded by 0 and 1. Using a link function, you derive a model using maximum likelihood. The result is $\log(p / (1 - p)) = 1.233 + 6.39$ x prospect's age and $p = 1/(1 + e^{-(1.23 + 6.39 x prospect's age)})$.

The formula enables you to estimate the response rate, given the value for *prospect's age*.

This example is a simple one; logistic regression models typically have numerous predictive variables.

## Neural Networks

Neural network processing differs greatly from regression in that the process does not follow any statistical distribution. Instead, a neural network is modeled after the function of the human brain. The process is one of pattern recognition and error minimization. You can think of it as ingesting information and learning from each experience.

Neural networks are made up of nodes that are arranged in layers. This construction varies, depending on the type and complexity of the neural network. Figure 3.9 illustrates a simple neural network with one hidden layer. Before the process begins, the data is split into *training* and *testing* data sets. (A third group is sometimes held out for final validation.) Then weights or "inputs" are assigned to each of the nodes in the first layer. During each iteration, the inputs are processed through the system and compared with the actual value. The error is measured and fed back through the system to adjust the weights. In most cases, the weights improve in their ability to predict the actual values. The process ends when a predetermined minimum error level is reached.

**Figure 3.9: Neural Network Schema Showing Flow of Data**

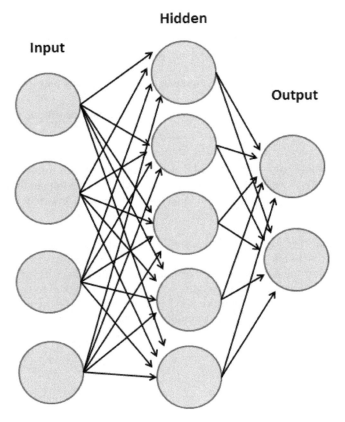

One specific type of neural network commonly used in marketing uses sigmoidal functions to fit each node. Recall that the sigmoidal function is the same function used in logistic regression. You might think about this type of neural network as a series of "nested" logistic regressions. This technique is powerful in fitting a binary, or two-level, outcome, such as a response to an offer or a default on a loan.

One of the advantages of a neural network is its ability to pick up nonlinear relationships in the data. This ability to pick up nonlinear relationships allows users to fit some types of data that would be difficult to fit by using regression. One drawback, however, is its tendency to overfit the data. Overfitting can cause the model to deteriorate more quickly when applied to new data. If this is the method of choice, be sure to carefully validate your data. Another disadvantage to consider is that the results of a neural network are often difficult to interpret.

# Modeling Process

The modeling process has four main stages:

1. Define the objective.
2. Develop the model.
3. Implement the model.
4. Track and monitor the model.

## Define the Objective

Defining the objective is the main theme of Chapter 1 because it is the most important step in the model building process. You can develop a perfect model using perfect data. But if it doesn't meet your business objective, it is a waste of time and money. Make sure the objective of the model is in alignment with your overall business goals.

## Develop the Model

The model development process comprises five steps.

1. Select the data.
2. Explore the data.
3. Modify the data and variables.
4. Build the model.
5. Assess the model.

If you follow this sequence, you will likely be successful in your model-building efforts.

### Step 1: Select the Data

Once you have defined your objective, you need to get data for developing your model. Be sure it correlates with the business purpose for building the model. The data can come from a variety of sources. In some cases, data will need to be combined. The specific details of the data selection are highly specific to the project. To learn about many sources for data construction, see Chapter 4.

### Step 2: Explore the Data

Once your data is available for analysis, you can use visual tools to view the distribution and completeness of the data. Several tables and graphs can display the data values and show you where data is incomplete or missing.

### Step 3: Modify the Data and Variables

In this step, you can partition your data into training, test, and validation data sets. On the basis of your findings in your data exploration step, you can impute missing values and filter outliers.

To ensure the best fit for regression and neural network models, you may want to transform or segment continuous variables. To use class variables in a regression or neural network model, you need to make them numeric. The best way to do so is to create indicator variables for any variables with character or non-sequential values. For an example of this technique, see Chapter 9.

For continuous variables, you can select certain transformations to provide a better fit for your model. SAS Enterprise Miner gives you the option of letting the software pick the best transformation, as is demonstrated in Chapter 9. For more detail on variable transformations, see *Data Mining Cookbook* (Parr-Rud 2001, 85–97).

## Step 4: Build the Model

The process of building the model is greatly streamlined within SAS Enterprise Miner. But it is still good to understand the underlying processes so that you are well prepared to uncover anomalies and errors.

### Selection Method

Earlier in the chapter, you learned about three modeling techniques: decision tree, regression, and neural networks. Each of these techniques offers unique considerations when you are determining your selection method.

For the decision tree model, the variables are split into discrete groups; therefore, you don't have to do any variable transformations or define a selection process. You can interactively build a tree if you are looking to select the variables that you want to enter the model. For information about building a tree interactively, see *Decision Trees for Analytics Using SAS Enterprise Miner* (DeVille and Neville 2013).

The order in which the variables are selected may affect the final outcome of the model. Many selection methods exist. Each one has its advantages and disadvantages. Depending on your particular business problem or resource situation, one method might be preferred over another.

For more detail on selection options for regression, see *Data Mining Cookbook* (Parr-Rud 2001, 104–105).

### Optimal Number of Variables

A common question in relation to the model-building process is, what is the optimal number of variables?

The answer depends on the situation. Most models that predict well have 7 to 15 variables. In certain situations, using every variable might be optimal. Although the model might be large and difficult to interpret, the extra variables do not detract from the power or fit of the model. In some businesses where analytic resources are scarce and CPU power is abundant, using every variable is often the best option.

## Multicollinearity

Another common question raised among predictive modelers is, should I worry about multicollinearity?

*Multicollinearity* exists when two or more independent, or predictive, variables are correlated with each other. This situation makes interpretation of the model equation difficult and is a common occurrence in scientific and medical research. However, when you develop a model for use in marketing or risk management, the interpretation of the model equation is not necessary for the model to be useful. So if the models are slightly correlated, it is not a problem. Moreover, if you use the sequential selection methods described in Step 4, then the correlation of the variables will be minimized. For more information on selection options for regression, see *Data Mining Cookbook* (Parr-Rud 2001, 106–108).

## Step 5: Assess the Model

At this point, you are ready to evaluate your model. The main tool for evaluating the power of the model is the decile and lift analyses.

## Gains Table

The gains table is a favorite among marketers and risk managers for evaluating and comparing models (Table 3.4). The most common type of assessment method in marketing and risk is decile analysis, which looks at the model performance in 10% groupings, or *deciles*. Within each decile, the gains table compares the model performance to the expected result that would have occurred if the names had been selected randomly–in essence, without the benefit of a model. The random result can also be referred to as the *average performance*.

**Table 3.4: Gains Table Displaying Deciles Measure of Model Performance**

| Decile | Number of Records | Mean Predicted Response (%) | Mean Actual Response (%) | Cumulative Mean Actual Response (%) | Number of Responders (% of Total) | Cumulative Total Responders (%) | Lift | Cumulative Lift |
|--------|-------------------|------------------------------|---------------------------|--------------------------------------|------------------------------------|----------------------------------|------|-----------------|
| 1 | 8,468 | 7.52 | 7.53 | 7.53 | 638 (24.08) | 24.08 | 241 | 241 |
| 2 | 8,467 | 5.08 | 5.04 | 6.29 | 427 (16.12) | 40.20 | 161 | 201 |
| 3 | 8,466 | 4.00 | 4.44 | 5.67 | 376 (14.19) | 54.40 | 142 | 181 |
| 4 | 8,467 | 3.19 | 2.92 | 4.98 | 247 (09.32) | 63.72 | 93 | 159 |
| 5 | 8,463 | 2.65 | 2.40 | 4.47 | 203 (07.66) | 71.39 | 77 | 143 |
| 6 | 8,472 | 2.29 | 2.03 | 4.06 | 172 (06.49) | 77.88 | 65 | 130 |
| 7 | 8,466 | 2.02 | 2.08 | 3.78 | 176 (06.64) | 84.52 | 66 | 121 |
| 8 | 8,467 | 1.79 | 1.62 | 3.51 | 137 (05.17) | 89.69 | 52 | 112 |
| 9 | 8,467 | 1.59 | 1.62 | 3.30 | 137 (05.17) | 94.87 | 52 | 105 |
| 10 | 8,476 | 1.36 | 1.60 | 3.13 | 136 (05.13) | 100.00 | 51 | 100 |
| Total | 84,679 | NA | 3.13 | NA | 2,649 (100.00) | NA | NA | NA |

You create the deciles by sorting the file by its score or probability (high to low) and by then dividing it into 10% groupings. Each decile can display a number of calculations including quantities, mean performance, and lift.

You see that Table 3.4 offers a lot of information. The following list describes each column in order, from left to right:

1. *Decile:* Ordered from 1 to 10, decile 1 is the best performing decile and decile 10 is the worst performing decile based on response rate.
2. *Number of Records*: The value for records is the number of people or other data points in the population. Each decile is approximately 10% of the population.
3. *Mean Predicted Response (%):* Column 3 displays the average response rate predicted by the model.
4. *Mean Actual Response (%)*: Column 4 displays the average actual response rate reflected in the data.
5. *Cumulative Mean Actual Response (%)*: The average actual response rate is reflected in the data cumulative down through each decile.
6. *Number of Responders (% of Total)*: Column 6 displays the count of responders in each decile.
7. *Cumulative Total Responders (%)*: Column 7 displays the cumulative count of responders down through each decile.
8. *Lift*: Column 8 displays the model efficiency at each decile (further explanation below).
9. *Cumulative Lift*: Column 9 displays the cumulative model efficiency down through each decile (further explanation below).

*Lift* is a measure of the model's ability to beat the random approach or average performance. *Lift* is another way of describing the model's efficiency. The higher the *lift*, the more powerful the model. The *lift* in Decile 1 is 241. This means that the model performance is 2.41 times better than random or average in the top decile.

*Cumulative lift* shows the efficiency of a group of deciles starting with the best. For example, if we look at the top three deciles: if you mailed a random 30% of the file, you would expect to get 30% of the potential responders. Looking at column 6 in the top three deciles, the model captures $(638 + 427 + 376)$ / $2649 = 54.40\%$ of the potential responses. So the *cumulative lift* is $54.40/30*100 = 181$. This result is equivalent to saying that you get an 81% higher-than-average response rate at a 30% depth of file, or that the model does 1.81 times better than average at 30% depth of file.

## Gains Chart

A *gains chart* is a graphical representation of the cumulative percentage of a target group as compared to a random or average result at each decile, or 10% of the file. It is an excellent visual tool for showing the power of a model (Figure 3.10). For example, at 50% of the file, you can capture 71% of the responders, which represents a 45% increase in responses over a random selection.

**Figure 3.10: Gains Chart Comparing Percentage of Responders for the Model with Percentage of Random Selection of the General Population**

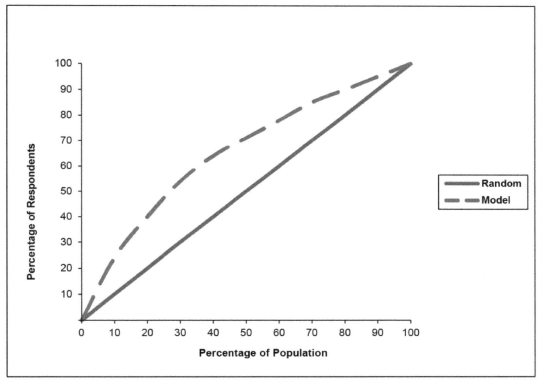

## *C* **Statistic**

The *C* statistic is useful in providing a quick assessment of the model fit. It is a function of the percent concordant, percent discordant, percent tied, and total pairs. They are defined as follows:

- *Percent Concordant:* For all pairs of observations with different values of the dependent variable (response = 1/0), a pair is *concordant* if the observation with target value (response = 1) has a higher predicted event probability than the observation with the nontarget value (response = 0).
- *Percent Discordant:* For all pairs of observations with different values of the dependent variable (response = 1/0), a pair is *discordant* if the observation with target value (response = 1) has a lower predicted event probability than the observation with the nontarget value (response = 0).
- *Percent Tied:* For all pairs of observations with different values of the dependent variable (response = 1/0), a pair is *tied* if the observation with target value (response = 1) has an *equal* predicted event

probability to the observation with the nontarget value (response = 0). In other words, it is neither concordant nor discordant.

- *Pairs:* This number is the total of possible paired combinations with different values of the dependent variable. The *C* statistic is defined as $(nc + 0.5(t - nc - nd))/t$, where $t$ = the total number of pairs with different response values, $nc$ = the number of concordant pairs, and $nd$ = the number of discordant pairs.

---

## Implement the Model

Model implementation is not a step in the model-building process per se, but discussing how to implement the model correctly is important for one reason: a perfectly good model will look like a bad model if not implemented correctly.

In many situations, a model is developed to replace an existing model. The old model might not be performing, or perhaps some new predictive information is available that can be incorporated into a new model. Whatever the reason, you will want to compare the new model, or the "challenger," to the existing model, or "champion." Again, depending on your goals, various ways exist to select your group of names for testing.

In Figure 3.11, the entire population is represented by the rectangle. The ovals represent the names selected by each model. If your "champion" model is still performing well, you might decide to mail the entire set of names selected by the "champion" and to mail a sample of names from the portion of the "challenger" oval that was *not* selected by the "champion." This technique allows you to weight the names from the sample so that you can track and compare both models' performance. If you want to estimate the performance of the population without the benefit of any model (for comparison purposes), you should mail a sample of names not selected by either model.

**Figure 3.11: Comparison of Champion with Challenger in Prospect Selection**

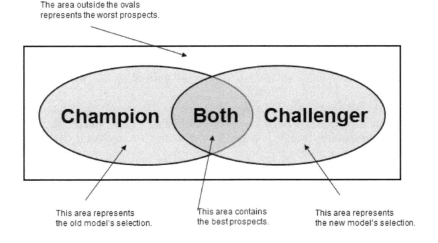

## Maintain the Model

At this point, you have invested much time and many resources in creating a powerful model. Now you can ensure its longevity and usefulness by documenting and tracking it. Since the mid-1990s, I have worked with numerous modelers, marketers, and managers to develop and implement predictive models. Shockingly, often little is known about what models exist within the corporation, how they were built, and how they have been used to date. After all the time and effort spent developing and validating a model, the extra effort to document and track the model's origin and utilization is certainly worthwhile. You begin by determining the expected life of the model.

## Model Life

The life of a model depends on several factors. One of the main factors is the target. If you are modeling response, you can redevelop the model within a few months. If the target is risk, then you may not know how the model performs for a couple of years. If the model has an expected life of several years, then you may be able to track the performance along the way.

Some predictive models are developed on data with performance appended. If the performance window is three years, it should contain all the activity for the three-year period. In other words, if you want to predict bankruptcy over a three-year period, you will take all names that are current for time, $T$. The performance will then be measured in the time period from $T + 6$ to $T + 36$ months. So when the model is implemented on a new file, the performance can be measured or benchmarked at each six-month period.

If the model is not performing as expected, then the choice has to be made whether to continue use, rebuild, or refresh.

When a model begins to degrade, the decision must be made to rebuild or refresh the model. To *rebuild* means to start from scratch. You would use new data, build new variables, and rework the entire process. To *refresh* means to keep the current variables and rerun the model on new data.

Refreshing the model usually makes sense, unless an opportunity emerges to introduce new predictive information. For example, if a new data source becomes available, then incorporating that information into a new model might make sense. If a model is very old, testing the building of a new one might be advisable. Finally, if strong shifts occur in the marketplace, then a full-scale model redevelopment may be warranted, as happened in the credit card industry when low introductory rates were launched. In that event, the key drivers for response and balance transfers were changing with the drop-in rates.

## Model Log

A model log is a record that, for each model, contains information such as development details, key features, and an implementation log. It tracks models over the long term with details such as the following:

- *Model Name*: Select a name that reflects the objective or product. Combining it with a number allows for tracking redevelopment models. Note the version number if applicable.
- *Time of Model Development*: Documenting the time range in which the model is developed allows you to track the entire model development process in relation to general business activities. This can be helpful when you are tracking hundreds of models within your organization.
- *Model Developer:* Because model development is as much an art as a science, models can have certain characteristics based on the person who developed the model.
- *Overall Objective.* The reason or purpose of the model is critical for ongoing relevance and tracking.
- *Model Type:* The technique used to build the model can include linear regression, logistic regression, decision tree, neural network, or ensemble models.
- *Specific Target:* The specific group of interest or value estimated assists in model deployment and comparison.
- *Model Development Campaign Data:* The campaign data used for model development identifies model details based on campaign characteristics.
- *Model Implementation Campaign Data:* The campaign data used for the first model implementation identifies model details based on campaign characteristics.
- *Model Implementation Launch Date:* Store the first date of the model implementation campaign.
- *Score Distribution (Validation):* The mean, standard deviation, minimum and maximum values of score characterize the structure of the validation sample.
- *Score Distribution (Implementation):* The mean, standard deviation, minimum and maximum values of score characterize the structure of the implementation sample.
- *Selection Criteria:* The score cut-off or depth of file represents the selection criteria
- *Selection Business Logic:* The business justification for the selection criteria helps define the model.
- *Pre-Selects:* The cuts prior to scoring are essential for accurate model implementation.
- *Expected Performance:* Note the expected rate of the target variable, such as response, approval, or other targeted action, which matches the specific target of the model.
- *Actual Performance:* Compare the actual rate of the target variable—such as response, approval, or other targeted action—to the expected performance.
- *Model Details:* The model details are characteristics about the model development that might be unique or unusual.
- *Key Drivers:* Key predictors in the model are the variables with the most power in the model.

A model log saves hours of time and effort because it serves as a quick reference for managers, marketers, and analysts to see what's available, how models were developed, who the target audience is, and more.

For best results, use a spreadsheet with a separate sheet for each model. One sheet might look something like Table 3.5. A new sheet should be added each time a model is used. Each sheet should include the all of the elements in the log. If one model is a modification of another model, that can be noted in the version number.

**Table 3.5: Sample Model Log as Tool for Organizing Important Tracking Details**

| Log Item | Detail |
|---|---|
| Name of Model | R2000 V1 |
| Time of Model Development | MM/YY–MM/YY |
| Model Developer | Name |
| Overall Objective | Increase response |
| Type of Model | Logistic regression |
| Specific Target | Responders with > $10 purchase |
| Model Development Campaign (Date) | Spring Campaign (MM/YY) |
| First Campaign Implementation (Date) | Fall Campaign MM/YY |
| Model Implementation Launch Date | MM/DD/YY |
| Score Distribution (Validation) | Mean = .037, SD = .00059, Min = .0001, Max = .683 |
| Score Distribution (Implementation) | Mean = .034, SD = .00085, Min = .0001, Max = .462 |
| Selection Criteria | Deciles 1 through 5 |
| Selection Business Logic | Expected response > .06 |
| Preselects | Age 25-65; Minimum risk screening |
| Expected Performance | $25,500 |
| Actual Performance | $26,243 |
| Execution Details | Sampled lower deciles for model validation and redevelopment |
| Key Drivers | Date of last purchase, population density |

This document will save you time and prevent errors and duplicate efforts. When combined with other model logs, it can be used as a reference library for all modeling projects within your department or company.

## Notes from the Field

Big data is a driving force in business today. It's only getting bigger, so the pressure to use big data to gain or maintain a competitive advantage is only going to increase. This trend and the complexity it entails make the right kind of analysis critical.

When you are first introduced to a data set or you are given a new business challenge or opportunity, it's tempting to leap ahead and build that fancy predictive model. But you should always begin by really understanding the underlying data patterns and how the different data characteristics relate to one another. If your goal is to develop a predictive model, then the initial insights gleaned from the descriptive analysis will provide clues for building your variables and data interactions. Descriptive analysis may also alert you to underlying data problems, like missing values and data biases, earlier in the process. If you are working with a client or team, you may get some benefit from sharing your early findings from the descriptive analysis with stakeholders who may have additional business insights or suggestions.

# Chapter 4: Data Construction for Analysis

## Introduction

Selecting the best data for analysis requires a thorough understanding of your business objective and your market. The data serves as the foundation for all of your analysis. In this case, the old adage is true: "Garbage in, garbage out!"

Finding the right data for analysis typically begins with your own data. Depending on your goal, you may have all the data you need in your own databases. However, in certain situations, you may need to purchase data from outside data sources to enhance your analysis. Either of these scenarios requires a culture of big data–a culture that supports collaboration and data sharing between silos and across systems. This culture of big data requires a strategic approach to data acquisition, storage, management, and access.

In this chapter, we will look at some typical sources for descriptive and predictive analysis. These examples give you a basic framework from which to develop your own analysis data sets.

## Data for Descriptive Analysis

As we saw in Chapter 3, *descriptive analysis* helps you understand the underlying characteristics and patterns in your data. Your data can come from any source. Descriptive analysis can analyze a point in time or can include trending of data across defined time intervals. Some kinds of descriptive analysis are as follows:

- *Customer profiling.* To get a good understanding of your customers, you can perform this analysis on your customer database. It involves looking at the different levels of categorical variables and the distributions of continuous variables.
- *Marketing activity.* Campaign analysis or any variation of marketing activity can come from the marketing database. It can be matched to the customer database to look at marketing performance by different customer segments.
- *Performance analysis.* Similar to marketing activity, performance analysis looks at your company's measures of success at a point in time. Having several "snapshots" at different points in time allows for trending. And this data can be matched to the customer database for analysis by customer segment.

## Data for Predictive Analysis

Data for predictive analysis generally comes from some historical or static source. Most commonly, a group of names is extracted or purchased with a group of predictive variables from a specified time period. The names are processed for action during a processing timeframe. Next, an offer or request is made to a group of customers or prospects. Each customer or prospect can respond or act on the stimulus during a fixed time period, often called the *performance window*.

At the end of the performance window, the responders or those who act on the stimulus are matched back to the original file from the earlier time period. The matched data forms the *modeling data set.* The model is then developed and scored for future use. This description is the basic framework for a modeling data set. Many examples will illustrate this process throughout the rest of this chapter.

As seen in Figure 4.1, a general timeline exists for the development of model data. The timeline for these steps can vary depending on the type of campaign or the channel used. For a direct-mail model, the historic data might be compiled for six months or more. The processing might take a month and the performance window might last three months. The model development might take anywhere from two to six weeks.

**Figure 4.1: General Timeline for Development of Model Data**

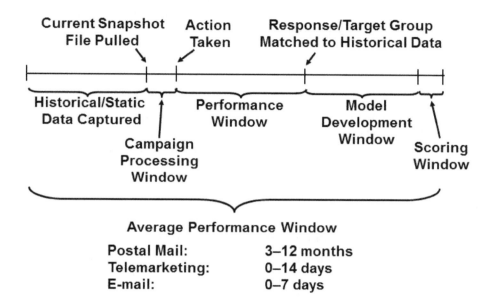

A telemarketing scenario might also use six months of data, or it might use results from the previous day. What is used varies greatly, depending on the situation. Processing can be anywhere from a few hours to a week. The performance window might be limited to the time spent on the phone. Development of the model might take as little time as a few hours to as much as a few weeks.

Internet models are becoming so sophisticated that companies are embedding automated modeling tools to capture historical data (which could represent the most recent 10 minutes of someone's online activity), develop models in real time, and generate offers or take action automatically via either email or postal mail.

## Prospect Models

The first kind of data discussed here is data for modeling prospects. This data is used for prospecting or acquiring new customers.

Data from a prior campaign is the best option for target modeling. This advice is true, whether or not the prior campaign matches the exact product or service you are modeling. Campaigns that have been generated by your company will be sensitive to factors like creative copy, terms of the offer or promotion, and brand identity. These factors may have a subtle effect on model performance.

It's important to note that data from a prior campaign for the same product and to the same group is the best option for data in any targeting model. It allows for the most accurate prediction of future behavior. The only factors that can't be captured in this scenario are seasonality, changes in the marketplace, and the effects of multiple offers.

If data from a prior campaign is not available, the next best option is to build a *propensity model*. To do so, you use a data source to develop a model that targets a product or service similar to your primary targeting goal.

Increasingly, companies are forming affinity relationships with other companies to pool resources and increase profits. For example, credit card banks are forming partnerships with airlines, universities, clubs, retailers, and energy companies. Telecommunications companies are forming alliances with airlines, insurance companies, and others. One of the primary benefits of such arrangements is shared access to personal information that can be used to develop targeting models.

The following scenarios highlight some common ways that modeling data sets are created.

## Case 1: Same Product to the Same List, Using a Prior Campaign

In January, ABC Credit Card Bank purchased 100,000 randomly selected names from Experian Credit Bureau for an acquisition campaign with the goal of developing a model. Together with the names, ABC purchased more than 300 demographic and credit attributes. ABC then mailed a credit card offer with an annualized annual percentage rate (APR) of 11.9% and no annual fee. ABC received approximately 1,000 responses from the campaign over the next 12 weeks. The responses were matched back to the original mail file to create a modeling data set.

In May, ABC Credit Card Bank built a response model, using the 300 variables that were purchased at the time of the original offer. After the model was constructed and validated, ABC used the model algorithm to score (assign a probability estimating their likelihood to respond) the May customer file at Experian Credit Bureau. In early June, it sent the same offer to the 30% of customers with the highest predicted response rate based on the model. To validate that the model performed as expected, it also mailed a random sample of the lower-scoring names.

## Case 2: Same Product to a New List, Using a Prior Campaign

Next, ABC Credit Card Bank wants to develop a response model for its standard 11.9% APR "No Annual Fee" offer, a model that will be optimal for scoring names at the Equifax Credit Bureau. All of the other terms and conditions are the same as the prior campaign that used Experian data. Rather than start with a random sample, ABC's most cost-effective option for getting data for model development is to use the model that was developed for the Experian Credit Bureau. ABC plans to use the Experian model to score the Equifax data. It will mail to the highest-scoring 40% of names selected by the model. To ensure a data set for developing a robust response model that is more accurate for the Equifax Credit Bureau, ABC will take a random sample of the names not selected by the model.

The response rate for the names in the top 40% is 2.1%. This rate is lower than the response rate from the Experian data. However, ABC was able to determine that the model still worked well; because the rest of the file was sampled, the performance for the entire file could be recreated by weighting the sample, which allows each member of the sample to be counted multiple times. The overall response rate was less than 1%, and as the score increased, each 10% increment of the population had a higher response rate. The recreated population file was used to develop a new model using only Equifax data.

## Case 3: Same Product to the Same List, Using Prior Campaign with Selection Criteria

Summit Sports is a company that sells sports equipment online and via catalog. Six months ago, Summit Sports purchased a list of prospects from Power List Company. The list contained names, email addresses, and 35 demographic and psychographic attributes. Summit Sports used criteria that selected only males aged 30 to 55 years. To that segment, it emailed a catalog that featured baseball products. After three months of performance activity, response and sales amounts were appended to the original email file to create a modeling data set.

Using the 35 demographic and psychographic attributes, Summit Sports plans to develop a predictive model to target responses with sales amounts that exceeded $20. After the model is constructed and validated, Summit Sports will have a robust tool for scoring a new set of names for targeting purchases exceeding $20 from the baseball products catalog. For best results, the names will be purchased from Power List Company and use the same selection criteria as were used for the first model.

## Case 4: Similar Product to Same List, Using Prior Campaign

A few months ago, RST Cruise Line decided to purchase a list of travel magazine subscribers from TravelPro Publishing Company. It randomly selected 10,000 names and sent them the Caribbean cruise offer. At the end of the performance window, when most of the expected responses were received, RST developed a model to use for targeting future Caribbean cruise offers.

Meanwhile, the product manager for the Alaskan cruise line wanted to try the same list of travel magazine subscribers. Rather than begin with a random sample, RST decided to use the Caribbean model to score the file for the Alaskan cruise offer. Even though the Alaskan cruise is less popular than the Caribbean one, it reasoned that at least, this way, it would target people who like to take cruises. By mailing the best 30% of the list and randomly sampling the rest of the list, RST validated that using the Caribbean model was more cost effective than just mailing a random offer.

## Customer Models

As markets mature in many industries, attracting new customers becomes increasingly difficult. For example, credit card banks are compelled to offer low rates to lure customers away from competitors. The cost of acquiring a new customer has become so expensive that many companies are expanding their product lines to maximize the value of existing customer relationships. Credit card banks are offering insurance or investment products, or they are merging with full-service banks and other financial institutions to offer a full suite of financial services. Telecommunications companies are expanding their product and service lines or merging with cable and Internet companies. Many companies in a variety of industries are viewing their customers as their key asset.

These expanding markets create many opportunities for target modeling. A customer who is already doing business with you is typically more likely to prefer you over a competitor when purchasing additional products and services. This likelihood creates many opportunities for cross-sell and up-sell target modeling. Retention and renewal models are also critical for targeting customers who may be looking to terminate their relationship. Simple steps to retain a customer can be quite cost-effective.

Data from prior campaigns is also the best data for developing models for customer targeting. Although most customer models are developed with internal data, overlay data from an external source is sometimes appended to customer data to enhance the predictive power of the targeting models.

> **TIP:** Many list companies will allow you to test their overlay data at no charge. If a list company is interested in building a relationship, it usually is willing to provide its full list of attributes for testing. The best method is to take a past campaign and overlay the entire list of attributes. Next, develop a model to learn which attributes are predictive for your product or service. If you find a few very powerful predictors, you can negotiate a price to purchase these attributes for future campaigns.

## Case 1: Cross-Sell

Sure Wire Communications has built a solid base of long distance customers over the past 10 years. It is now expanding into cable television and wants to cross-sell this service to its existing customer base. Through a phone survey to 200 customers, Sure Wire learned that approximately 25% are interested in signing up for cable service. To develop a model for targeting cable customers, it wants a campaign with a minimum of 5,000 responders. It is planning to mail an offer to a random sample of 25,000 customers. This plan will ensure that, with as low as a 20% response rate, it will have enough responders to develop a model.

## Case 2: Up-Sell Using Life-Stage Segments

XYZ Life Insurance Company wants to develop a model to target customers who are most likely to increase their life insurance coverage. From past experience and common sense, it knows that customers who are just starting a family are good candidates for increased coverage. But it also knows that other life events can trigger the need for more life insurance.

To enhance its customer file, XYZ plans to test overlay data from Lifetime List Company. Lifetime specializes in *life-stage* segmentation. XYZ feels that this additional segmentation will increase the power of its model. To improve the results of the campaign, XYZ Life is planning to make the offer to all of its customers in Life Stage 3. These customers are the ones who have a high probability of being in the process of beginning a family. XYZ Life will pull a random sample from the remainder of the names to complete the mailing. Once the results are final, it will have a full data set with life-stage enhancements for model development.

## Case 3: Loyalty, Attrition, and Churn

Float Credit Card Bank wants to predict which customers are going to pay off their balances in the next three months. Once they are identified, Float will perform a risk assessment to determine whether it can lower its APR, in an effort to keep its balances. Through analysis, Float has determined that there is some seasonality in balance behavior. For example, balances usually increase in September and October because of school shopping. They also rise in November and December because of holiday shopping. Balances almost always drop in January as customers pay off their December balances. Another decrease is typical in April, when customers receive their tax refunds.

To capture the effects of seasonality, Float decided to look at two years of data. It restricted the analysis to customers who were out of their introductory period by at least four months. The analysts at Float structured the data so that it could use the month as a predictor, together with all the behavioral and demographic characteristics on the account. The modeling data set was made up of all the lost customers and a random sample of the current customers.

## Risk Models

Managing risk is a critical component to maintaining profitability in many industries. Most of us are familiar with the common risk inherent in the banking and insurance industries. The primary risk in banking is the borrower's failure to repay a loan. In insurance, the primary risk is the customer's filing a claim. Another major risk assumed by banks and insurance companies is that of fraud. Stolen credit cards cost banks millions of dollars a year. Losses from fraudulent insurance claims are equally staggering.

Strong relationships have been identified between financial risk and some types of insurance risk. As a result, insurance companies use financial risk models to support their claim risk modeling. One interesting demonstration of the correlation between financial risk and insurance risk is the fact that credit payment behavior predicts the likelihood of auto insurance claims. Even though they seem unrelated, the two behaviors are clearly linked and are used effectively in risk assessment.

Risk models are challenging to develop for several reasons. The performance window has to cover a period of several years to be effective, which can make the models difficult to validate. Credit risk is sensitive to the health of the economy, and the risk of claims for insurance is vulnerable to population trends.

Credit data is easy to obtain from any of the three major credit bureaus: Experian, Equifax, and Trans Union. It's just expensive to request and can be used only for an offer of credit. Some insurance risk data, such as data from life and health insurance claims, is relatively easy to obtain, but obtaining claims data for the automotive insurance industry can be more difficult to obtain.

Because of the availability of credit data from the credit bureaus, you can build risk models on prospects. This availability of credit risk data creates quite an advantage to banks that are interested in developing their own proprietary risk scores.

## Case 1: Default Risk for Prospects

High Street Bank has been conservative in the past. Its product offerings were limited to checking accounts, savings accounts, and secured loans. As a means to attract new customers, it is interested in offering unsecured loans. But first, it wants to develop a predictive model to identify prospects that are likely to default.

To create the modeling and development data set, High Street Bank decides to purchase data from a credit bureau. It is interested in predicting the risk of bankruptcy for a prospect for a three-year period. The risk department requests 12,000 archived credit files from four years ago, 6,000 that show a bankruptcy in the last three years, and 6,000 with no bankruptcy. This combined file of names will give it a view of the customer's credit and other financial characteristics at that point in time.

## Case 2: Loss-Given-Default Risk for Prospects

Trustus Mortgage wants to estimate its exposure to loss before approving a new mortgage. Each customer already has a score that estimates the probably of default. The next step is to estimate the amount of the loss, given that the default occurs. The combined score provides a good estimate of the full level of risk.

The modeling data set is constructed by combining two samples. The first is a sample of current customers that have had their loan for at least five years. The second is a sample of customers whose loans were initiated at least five years ago and have completed the default process. The predictive information is built from the information on their original application. The loss information includes information related to the property, the location (to accommodate state and local laws), and the nature of the loan. For example, some states have different processes for foreclosure, some loans have mortgage insurance, and the local real estate market has an important role in the final amount of the loss.

## Case 3: Fraud Risk for Customers

Anchor Credit Card Bank wants to develop a model to predict fraud. It captures purchase activity for each customer in the transaction database, including the amount, date, and spending category. To develop a fraud model, it collects several weeks of purchase data for each customer and calculates the average daily spending within each category. From this information, it can establish rules that trigger an inquiry if a customer's spending pattern changes.

## Case 4: Insurance Risk for Customers

CCC Insurance Company wants to develop a model to predict comprehensive automobile claims for a one to four-year period. Until now, it has been using simple segmentation based on demographic variables from the customer database. It wants to improve its prediction by building a model with overlay data from Sure Target List Company. Sure Target sells demographic, psychographic, and proprietary segments called Sure Hits that it developed by using cluster analysis.

To build the file for overlay, CCC randomly selects 5,000 names from the customers with at least a five-year tenure who filed at least one claim in the previous four years. It randomly selects another 5000 customers with at least a five-year tenure who have never filed a claim. CCC sends the files to Sure Target List Company, with a request that the customers be matched to an archive file from five years ago. The demographic, psychographic, and proprietary segments represent the five-year-old customer profiles. The data can be used to develop a predictive model that will target customers who are likely to file a claim in the next four years.

> **TIP:** For best results in model development, strive to have the population from which the data is extracted resemble the population to be scored.

## External Sources of Data

Scores built on your own data from within your own company will almost always be the most powerful ones available. But, as shown in some of the examples above, data purchased from an outside source is often useful for testing and modeling. Some main sources of data for both consumers and businesses are the major credit bureaus Experian, Equifax, and Trans Union. Other companies that sell scores are Fair Isaacs (FICO), Dun and Bradstreet (business data), InfoUSA, and U.S. Data Corp. A little Internet research will uncover a multitude of data sources.

## Notes from the Field

As you can see, there are numerous sources and types of data available for modeling and analysis. With experience, you may find additional sources. Just be sure it is accurate and makes sense to your business. You learned in Chapter 2 that your analysis is only as good as your data, and this truism certainly applies to descriptive and predictive analyses. You can create a fancy analysis on bad data, but the analysis will be worthless. Similarly, you can build what appears to be a strong model on bad data, and the model will be worthless.

Once you understand the data and how it was sourced, the next step is to perform exploratory data analysis (EDA). You may be tempted to skip this step, but EDA is an important step in understanding the health of your data. It involves looking at frequencies of categorical variables and at distributions of continuous variables, also known as descriptive analysis (described in Chapter 3).

If you recall the Ferrari metaphor, SAS technology makes it possible for you to build a model without knowing whether the data is good or bad. So, take care to use common sense and business acumen to validate your results.

# Chapter 5: Descriptive Analysis Using SAS Enterprise Guide

## Introduction

In Chapter 1, you learned the importance of setting a clear objective. In Chapter 2, you looked at types and sources of data, using examples across a variety of industries. In Chapter 3, you were introduced to the concept of descriptive and predictive analysis. In Chapter 4, you learned how to prepare data for analysis. Now, you're ready to start making the data actionable. Beyond profiling and correlations, you can begin to craft analyses that improve your marketing actions or generate new product ideas. Exploring the data with a business goal in mind is one of the best ways to leverage the power of descriptive analysis.

Throughout this chapter, you will focus on a specific case study from the publishing industry. With slight modifications, this process can work in a variety of industries. So, while you work through this analysis, think about how you can adapt it for your own industry.

## Project Overview

The leadership team at DMR Publishing Company is interested in understanding the drivers of revenue within its business. For the past year, it has gathered U.S. customer data that consists of revenues, numbers of magazines or journals, and three demographic variables. For this analysis, you will use SAS Enterprise Guide 6.1.

The project has five steps:

1. Initiate the project in SAS Enterprise Guide 6.1.
2. Import and view the data.
3. Explore the data.
4. Segment and profile the data.
5. Perform correlation analysis.

## Project Initiation

Open the project (Figure 5.1) by double-clicking the SAS Enterprise Guide icon. Click **New Project**. You will see an area on the left called the **Project Tree**. The Project Tree is where you can track each process and output in the project. The **Process Flow** area on the right is where you will build your process, in the form of a flowchart.

**Figure 5.1: SAS Enterprise Guide 6.1 Work Area**

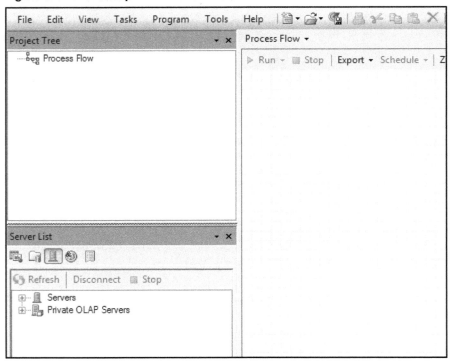

Because whole books on SAS Enterprise Guide basics are available, this book focuses on what is required for typical marketing analysis. For an excellent general reference, see *The Little SAS Book for Enterprise Guide* by Lora D. Delwiche and Susan J. Slaughter (2010).

# Exploratory Analysis

First, import the spreadsheet that contains the customer data.

## Importing the Data

Depending on your systems and storage, you may want to seek help from your IT department for access to the data. In our example, the DMR Publishing customer data is stored in the SASUSER library in the form of a Microsoft Excel spreadsheet. As shown in Figure 5.2, click **File ▶ Import Data**. When the input data window opens, click the up arrow to browse to the folder in which the spreadsheet, DMR Customer Base, is stored. Highlight the spreadsheet and click **Open**.

**Figure 5.2: Import Data into SAS Enterprise Guide**

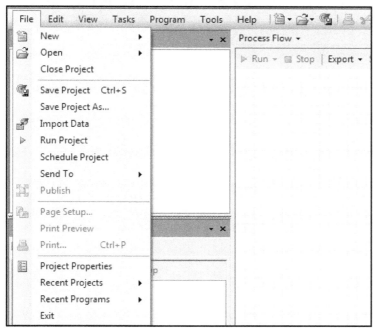

A menu will open to allow you to manage the import of the data. With regard to Steps 1 through 4, you are not adjusting anything in this process, so click **Finish**. Once the spreadsheet is imported, a window will open that displays the data set.

## Viewing the Data

The data appears to be in good condition and self-explanatory. Note that most data will require some cleaning. For information about this important process, see *Cody's Data Cleaning Techniques Using SAS* by Ron Cody (2008).

You will see seven variables, a few of which require definitions:

- The CUSTOMER_ID is a unique ID number that helps the company track the customer. You won't be using it in this analysis.
- CUSTOMER_SUBSCRIPTION_COUNT is the total number of current magazine and journal subscriptions held by the customer.
- CUSTOMER_REVENUE is the total revenue for the last year

**TIP:** Once the data is imported, you should save the project. Look for the icon on the top menu bar. The first time you do so, a menu will open and ask you to select a location and name the project. Once the project has been named, you can click that icon anytime, and the project will be saved under the same name.

## Exploring the Data

Once you have viewed the data to your satisfaction, close the data set by clicking the X in the upper right-hand corner of your screen. You can reopen it any time by right-clicking or double-clicking the data set icon that appears in both the Project Tree and the main work area.

Before you create reports, let's look at a helpful SAS Enterprise Guide option that allows you to send all output to a rich text format (RTF) file. Click **Tools ▶ Options ▶ Results**. When the Results section opens, you have several options for the format of the output data, as shown in Figure 5.3. The default is **SAS Report**. This default may be sufficient, but if you have Microsoft Word on your computer, you might click the box next to RTF near the top. This option allows you to share the documents in Microsoft Word format with all default settings that work well for most reports. You can also select **PDF**, **HTML**, or **Text Output**. You can explore these selections to see which works best for your projects.

**Figure 5.3: Format Options for Results**

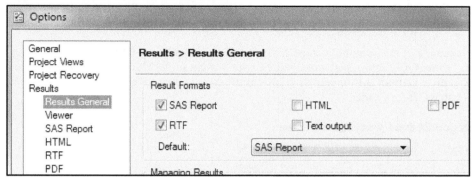

To view the data characteristics, you have several options. The fastest and simplest is to place your cursor on the data set in the work area and select **Tasks ▶ Describe ▶ Characterize Data** (Figure 5.4). This task is designed to give you a complete overview of every variable in your data set, in a useful format.

**Figure 5.4: Task Menu in SAS Enterprise Guide to Characterize Data**

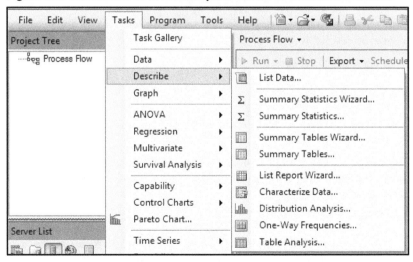

When the first menu appears, click **Finish.**

**NOTE:** If you have a categorical variable with more than 30 levels, such as *zip code*, Enterprise Guide will limit the output to the most frequent 30 levels. This option can be changed.

Once the process is complete, you can view the output by clicking on a tab at the top of the work area. If you have Microsoft Word on your computer, then you can access the RTF output by clicking the Microsoft Word icon. It will open the results in Microsoft Word outside of the project.

Output 5.1 is an example of the output of categorical variables in a frequency table. The results shown in Output 5.2 are fairly self-explanatory: males, with a count of 9,372, represent 59% of the sample; females, with a count of 6,140, represent 38% of the sample. Approximately 3% of the file has an unknown value (U) for gender.

**Output 5.1: Categorical Data Analysis of Gender**

| Variable | Label | Value | Frequency Count | Percent of Total Frequency |
|----------|-------|-------|-----------------|----------------------------|
| GENDER   |       | M     | 9372            | 58.5055                    |
|          |       | F     | 6140            | 38.3295                    |
|          |       | U     | 507             | 3.1650                     |

Output 5.2 provides results for continuous or interval variable AGE.

**Output 5.2 Interval or Continuous Data Analysis of Age**

| Variable | Label | N | NMiss | Total | Min | Mean | Median | Max | StdMean |
|----------|-------|-----|-------|--------|-----|-------|--------|-----|---------|
| AGE      |       | 16019 | 0 | 579193 | 18 | 36.16 | 34 | 92 | 0.10 |

For AGE, the following measures are available:

- *N* is the number of customers.
- *NMiss* is the number of missing values for Age. The value here is zero because missing values are represented by the value 'U' in our data set.
- *Total* equals the value if you sum age. It has no meaning in our analysis.
- *Min* and *Max* are the lowest value and the highest value.
- *Mean* equals the mathematical average or arithmetic mean (with missing value stripped).
- *Median* is the age of the customer at 50% of the file when sorted by age; it is the middle number when the observations are put in order.
- *StdMean* equals the standard error variation of the mean.

The graph in Output 5.3 shows the general distribution of the customer base by age. You can see that most of the customers are younger, with a heavy concentration between ages 21 and 40. In addition, there appear to be some outliers, or extreme values, represented by a few customers older than age 90. Outliers can cause problems with some types of analyses, especially regression models.

**Output 5.3: General Distribution of a Company's Customer Base by Age**

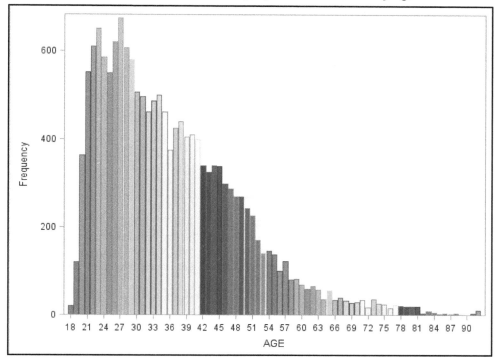

## Segmentation and Profile Analysis

The next step is to partition the customer base by CUSTOMER_REVENUE to compare the characteristics of the customers with the highest revenue to the rest of the customers.

Highlight the DRM Customer Base data set and select **Task ▶ Describe ▶ Distribution Analysis**. A window will open in which you can select the variables. Click CUSTOMER_REVENUE and move it to the area below Analysis Variables (Figure 5.5). In the left-hand column, select **Tables**. Check the box next to **Quantiles**. Click **Run**.

**Figure 5.5: Selection of Variable for Distribution Analysis of Customer Revenue**

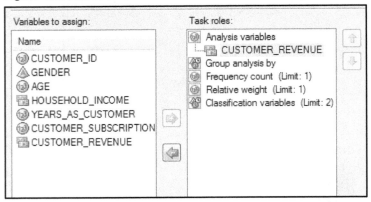

Output 5.4, captured from the RTF Output, shows the quantiles of the CUSTOMER_REVENUE distribution. A standard method that companies use to understand the drivers of revenue is to compare the characteristics of the top 25% of customers to the bottom 75%, on the basis of revenue. According to the Q3 value at the 75th percentile, anyone with more than $100 in revenue is in the top 25%.

**Output 5.4: Quantiles of a Customer-Based Revenue Distribution**

| Quantiles (Definition 5) | |
|---|---|
| Level | Quantile |
| 100% Max | 1040 |
| 99% | 260 |
| 95% | 180 |
| 90% | 140 |
| 75% Q3 | 100 |
| 50% Median | 60 |
| 25% Q1 | 30 |
| 10% | 15 |
| 5% | 15 |
| 1% | 15 |
| 0% Min | 15 |

Next, you will create a binary variable that will allow you to split the file between the top 25% and the bottom 75% for analysis.

To create a new variable, use **Advanced Expression** function within the **Query Builder**. Right-click the DMR Customer Base data set, then select **Query Builder** (Figure 5.6). Put your curser on the data set name, t1. (DMR_CUSTOMER_BASE, in the left-hand column), and drag it, together with all the variables, over to the right side under **Select Data.**

**Figure 5.6: Creation of a Binary Variable with Query Builder**

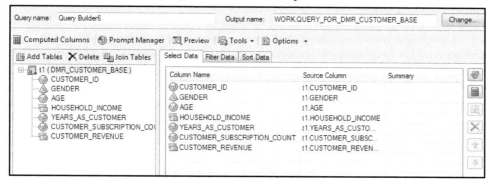

In the same window, click **Computed Columns**, then **New**, and finally, select **Advanced Expression** and click **Next.** A new window will open. In the empty box under **Enter an Expression**, type "case when CUSTOMER_REVENUE gt 100 then 1 else 0 end" as shown in Figure 5.7. Click **Next**.

**Figure 5.7: Use of Build Advanced Expression, CR_TOP_25PCT, within Query Builder**

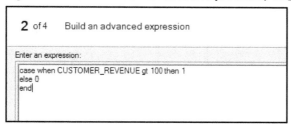

In window **3 of 4** of the Expression Builder, put CR_TOP_25PCT in the space next to **Column Name** (Figure 5.8) and click **Finish**.

**Figure 5.8: Assign Column Name to Computed Variable CR_TOP_25PCT**

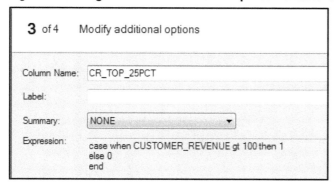

When this window closes, you are back in the Computed Columns window. Click **Close**. Complete the task by clicking **Run**.

CR_TOP_25PCT is your new binary variable that allows you to split the customer base between the top 25% and bottom 75% in revenue generation.

Now, you are ready to do the analysis. With your curser on the new data set, click **Tasks ▶ Describe ▶ Summary Statistics.**

When the window opens, slide AGE, HOUSEHOLD_INCOME, and CUSTOMER_REVENUE into the slots under **Analysis variables** (Figure 5.9). Slide the new binary variable CR_TOP_25PCT into the slot under **Classification variables**. Check the boxes next to **Mode** and **Sum**. Then, select **Percentiles** on the left and check the box next to **Median**. Click **Run**.

**Figure 5.9: Selection of Continuous Variables for Profiling**

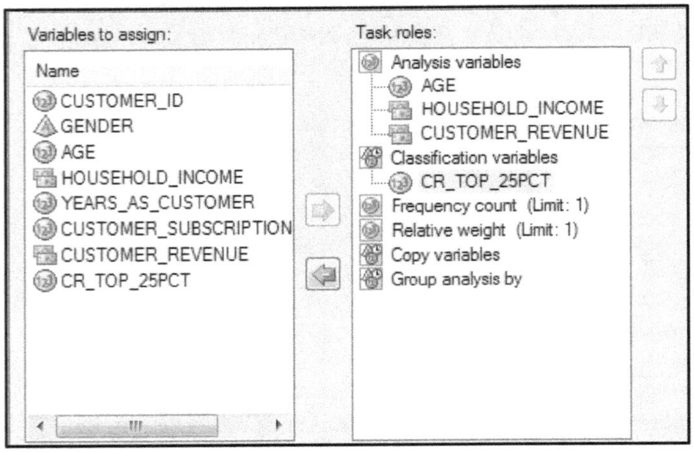

Next, look in the left-hand column and under **Data**, and click **Statistics** (Figure 5.10). Under **Basic**, on the right, set the **Maximum** decimal to 0. Uncheck the boxes next to **Minimum** and **Maximum**.

**Figure 5.10: Selection of Basic Statistics on Variables for Profiling**

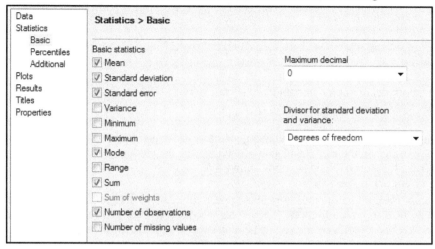

Next, select **Percentiles** on the left, and check the box next to **Median** (Figure 5.11). Click **Run**.

**Figure 5.11: Selection of Percentiles on Variables for Profiling**

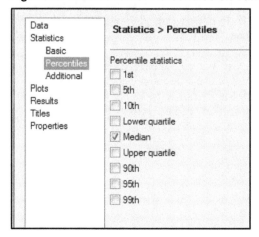

The resulting table offers much insight into what is driving profitability (Output 5.5).

**Output 5.5: Profiles for Continuous Variables by Revenue Group (Top 25% and Bottom 75%)**

| CR_TOP_25PCT | N Obs | Variable | Mean | Std Dev | Mode | Sum | N | Median |
|---|---|---|---|---|---|---|---|---|
| 0 | 12528 | AGE | 37 | 13 | 27 | 461415 | 12528 | 34 |
| | | HOUSEHOLD_INCOME | 49730 | 35213 | 36000 | 623015000 | 12528 | 44000 |
| | | CUSTOMER_REVENUE | 48 | 27 | 15 | 596735 | 12528 | 45 |
| 1 | 3491 | AGE | 34 | 12 | 20 | 117778 | 3491 | 31 |
| | | HOUSEHOLD_INCOME | 80307 | 49413 | 72000 | 280350000 | 3491 | 72000 |
| | | CUSTOMER_REVENUE | 161 | 93 | 120 | 561905 | 3491 | 140 |

According to the results, both demographic variables, AGE and HOUSEHOLD_INCOME, vary between the two groups:

- *Age*—The mean age of the top 25% of revenue generators is 34, compared with 37 for the bottom 75%. As shown on the right hand side of the display, the median varies by a similar amount at 31 and 34, respectively. The mode shows that the most common values for age are even lower at 27 and 20, respectively.

- *Household income*—Based on the mean, income appears to be more than 60% higher for the top 25%, with $80K as opposed to $49K. The median shows about the same amount of increase with $72K and $44K, respectively. The mode value of $72K for the top 25% is exactly double the mode value of $36K for the lower 75%.

- *Customer revenue*—The mean revenue per customer in the top 25% is $161 annually, as opposed to $48 annually from the bottom 75%. Another measure of interest comes from the sum. We want to know what percentage of the revenue is generated by the best 25% of the customer base. The result shows that the top 25% of customers bring in only slightly less revenue, $562K, than the bottom 75% with $597K. In other words, the best 25% of the customers generates almost half of the total revenue.

Gender can also be profiled to reveal whether a variation exists between the top 25% and the rest of the customer base. Because it is a categorical variable, you will use a frequency analysis. Place your cursor on the same data set you used for the Summary Statistics. Go to **Tasks ▶ Describe ▶ Table Analysis**. In the first window, drag GENDER and CR_TOP_25PCT into the boxes under **Table variables** as shown in Figure 5.12.

**Figure 5.12: Selection of Categorical Variables for Profiling Gender and the Top 25% of Customers Based on Revenue**

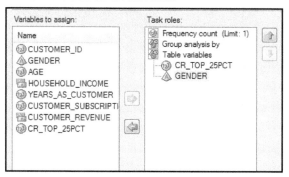

Next, click **Tables** in the upper left-hand menu (Figure 5.13). Drag CR_TOP_25PCT into the area to the left of the table. Drag GENDER into the area at the top of the table (see Figure 5.12). Click **Table Statistics ▶ Association,** then click **Chi-square tests.** Click **Run.**

**Figure 5.13: Construction of Table Showing Gender by the Top 25% of Customers Based on Revenue**

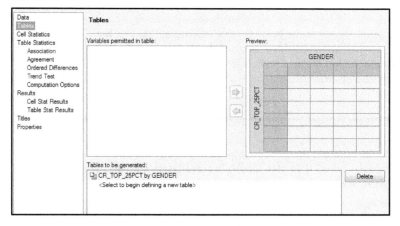

The results in Output 5.6 show that GENDER does differ between the top 25% and bottom 75% of customers based on revenue. A slightly greater number of females (F) are in the top 25% (24%, as opposed to 76%) than males (M; 21%, as opposed to 79%). The unknowns (U) show a ratio of male to female, but represent only a small number.

**Output 5.6: Profiles for Categorical Variables by Revenue Group (Top 25% and Bottom 75%)**

| Table of CR_TOP_25PCT by GENDER | | | | | |
|---|---|---|---|---|---|
| | | GENDER | | | |
| | | F | M | U | Total |
| CR_TOP_25PCT | | | | | |
| 0 | Frequency | 4682 | 7449 | 397 | 12528 |
| | Col Pct | 76.25 | 79.48 | 78.30 | |
| 1 | Frequency | 1458 | 1923 | 110 | 3491 |
| | Col Pct | 23.75 | 20.52 | 21.70 | |
| Total | Frequency | 6140 | 9372 | 507 | 16019 |

Statistics for Table of CR_TOP_25PCT by GENDER

| Statistic | DF | Value | Prob |
|---|---|---|---|
| Chi-Square | 2 | 22.6737 | <.0001 |
| Likelihood Ratio Chi-Square | 2 | 22.5213 | <.0001 |
| Mantel-Haenszel Chi-Square | 1 | 18.2288 | <.0001 |
| Phi Coefficient | | 0.0376 | |
| Contingency Coefficient | | 0.0376 | |
| Cramer's V | | 0.0376 | |

Sample Size = 16019

The level of statistical significance is shown in the bottom table of Output 5.6. For the **Chi-Square** statistic, $p < .0001$. Therefore, a statistically significant relationship exists between GENDER and CUSTOMER_REVENUE. The conclusion is that females are more likely than males and unknowns to be in the top 25% based on revenue.

# Correlation Analysis

Another simple but powerful way to examine the relationship between revenue and the continuous demographic variables (AGE and HOUSEHOLD_INCOME) is correlation analysis. This technique enables you to look at CUSTOMER_REVENUE as a continuous variable.

With the cursor on the data set, click **Tasks ▶ Multivariate ▶ Correlations**. In the window that will open, drag AGE, HOUSEHOLD_INCOME, and YEARS_AS_CUSTOMER into the area on the right under **Analysis Variables** (Figure 5.14). Drag CUSTOMER_REVENUE under **Correlate with**. Click **Run**.

**Figure 5.14: Selection of Continuous Variables for Correlation Analysis**

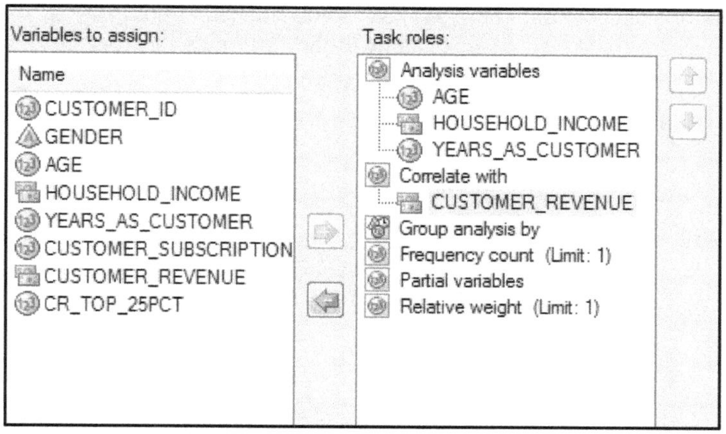

The results show that CUSTOMER_REVENUE has significant positive correlation for HOUSEHOLD INCOME, with a Pearson correlation coefficient of .38 (Output 5.7).

The *p* values for AGE and YEARS_AS_CUSTOMER also show significance. However, the Pearson correlation coefficients for these variables are each less than .10. Therefore, the significance is mainly due to the large sample size.

**Output 5.7: Results for Correlation Analysis of Customer Revenue with Age, Household Income, and Years as Customer**

| | AGE | HOUSEHOLD_INCOME | YEARS_AS_CUSTOMER |
|---|---|---|---|
| Pearson Correlation Coefficients, N = 16019 <br> Prob > \|r\| under H0: Rho=0 | | | |
| CUSTOMER_REVENUE | -0.08809 <br> <.0001 | 0.38130 <br> <.0001 | -0.08779 <br> <.0001 |

## Notes from the Field

Viewing data frequencies and distributions gives you a view of the data that is useful for understanding the underlying trends. Correlation analysis is powerful for detecting early relationships that can lead to ideas for predictive analysis.

Correlation analysis can also be helpful in detecting data problems. For example, if two variables show a high correlation coefficient (> 60%), they might be related in some way not otherwise known.

When descriptive analysis is the first step in a bigger project, you might want to discuss your preliminary findings with your client or stakeholders. They may have knowledge about or insights into market trends, potential data issues, and correlations that can be useful for you to incorporate in your analysis.

# Chapter 6: Market Analysis Using SAS Enterprise Guide

## Introduction

Chapters 1 through 5 focused on general knowledge and techniques that laid the foundation for many types of data analysis. In this chapter, you will explore a more specific topic: competitive analysis.

When your company sets its strategy, the first questions to answer are as follows:

- What are the strengths and weaknesses of our primary competitors as compared to us?
- What is our market share and how can we increase it?"

For most industries, data sources are available that allow companies to determine their market share, or "share of wallet," within certain segments of the population. This analysis is valuable for setting marketing strategies, guiding research and development, and informing finance and budget allocations.

## Project Overview

The leadership team at DMR Publishing Company is interested in understanding the drivers of revenue within its business. It has gathered U.S. customer data for the past year that consists of revenues, numbers of publications, and three demographic variables. For this analysis, you will use SAS Enterprise Guide.

The project has eight steps:

1. Initiate the project in SAS Enterprise Guide 6.1.
2. Import and view market data.
3. Add the DMR Publishing customer SAS data set to the project.
4. Use the query tool to build a new age group variable.
5. Summarize the DMR Publishing customer data to a gender and age group level.
6. Merge customer and market data.
7. Summarize the merged data to age group level.
8. Perform penetration and a "share of wallet" analysis.

## Market Analysis

So far in this analysis, our main interest has been to describe the characteristics of the customer base. If possible, it can be useful to compare characteristics of the customer database with the same characteristics in the overall market. This comparison enables a penetration analysis and share of wallet analysis as defined in Chapter 1.

We have purchased a data set from General List Company that shows the annual revenues and number of subscriptions for the entire publishing market within the United States, segmented by gender and age group. Our goal is to compare performance within gender and age group between the DMR Publishing's customer data and the total market data purchased from General List Company.

### Project Initiation

Similar to the process in Chapter 5, the first step is to open the project. Double click the SAS Enterprise Guide icon. Select **New Project**.

### Data Preparation

Your first step is to prepare the data.

#### Import Data

To access the data, go to **File ▶ Import** and choose the Publishing Market data set. When the first window opens, click **Finish**. Notice that the data is already summarized as shown in Output 6.1.

## Output 6.1: Publishing Market Data

| | SUBSCRIBERS | GENDER | AGE_GROUP | REVENUE | SUBSCRIPTIONS |
|---|---|---|---|---|---|
| 1 | 10,484 | F | 18-21 Years Old | $431,319 | 24,187 |
| 2 | 14,169 | F | 22-25 Years Old | $885,233 | 43,451 |
| 3 | 11,753 | F | 26-30 Years Old | $904,011 | 37,418 |
| 4 | 16,077 | F | 31-40 Years Old | $1,303,178 | 51,428 |
| 5 | 11,421 | F | 41-50 Years Old | $1,103,420 | 40,624 |
| 6 | 9,432 | F | 51+ Years Old | $781,581 | 30,789 |
| 7 | 8,561 | M | 18-21 Years Old | $289,773 | 16,728 |
| 8 | 15,836 | M | 22-25 Years Old | $829,967 | 42,518 |
| 9 | 16,105 | M | 26-30 Years Old | $1,091,799 | 47,651 |
| 10 | 22,319 | M | 31-40 Years Old | $1,806,043 | 69,480 |
| 11 | 15,180 | M | 41-50 Years Old | $1,458,032 | 53,122 |
| 12 | 12,288 | M | 51+ Years Old | $1,156,802 | 42,818 |
| 13 | 1,243 | U | 18-21 Years Old | $45,998 | 2,609 |
| 14 | 1,566 | U | 22-25 Years Old | $87,298 | 4,398 |
| 15 | 1,174 | U | 26-30 Years Old | $76,911 | 3,411 |
| 16 | 1,149 | U | 31-40 Years Old | $83,765 | 3,497 |
| 17 | 696 | U | 41-50 Years Old | $59,602 | 2,258 |
| 18 | 571 | U | 51+ Years Old | $42,800 | 1,728 |

To create the penetration and wallet analyses, merge the market data to the DMR Publishing customer data. Before you can merge the two data sets, you need to summarize the customer data. We plan to merge the two data sets by gender and age group.

The GENDER field is ready to match, but you must create the AGE_GROUP variable in the DMR Publishing customer data.

Because we started a new project, we need to bring the DMR Publishing customer data into the project. In the top menu, select **View ▶ Server List.** A new window will open below the Project Tree window; it will display the servers, as shown in Figure 6.1.

## Figure 6.1: Access Server List to Locate DMR Publishing Customer Data

Click the plus sign to the left of **Servers.** Continue to expand until you reach the **SASUSER library** as shown in Figure 6.2.

**Figure 6.2: Expand Library to Locate DMR Publishing Customer Data**

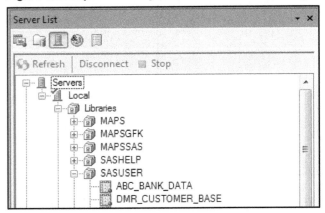

Right-click the DMR_CUSTOMER_BASE data and select **Open**. The data has now been added to the project and is visible in the Project Tree. You are ready to create the Age Group variable in the DMR Publishing customer data.

## Build Age Group Variable

Right-click the DMR Customer data set and select **Query Builder** as shown in Figure 6.3. Drag four variables from the left over to the right column: GENDER, CUSTOMER_ID, CUSTOMER_SUBSCRIPTION_COUNT, and CUSTOMER_REVENUE.

**Figure 6.3: Query to Build Age Group on DMR Publishing Customer Data**

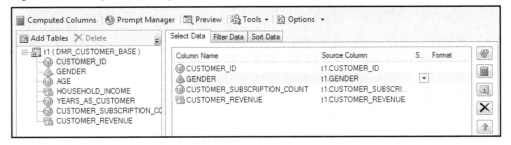

Next, click **Computed Columns** ▶ **New** ▶ **Advanced Expression** ▶ **Next** (Figure 6.4).

**Figure 6.4: Expression Builder to Create Age Group Variable**

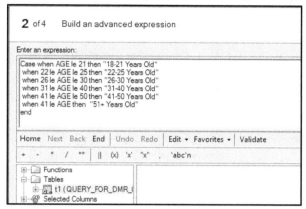

The window has a space to enter an expression. Program 6.1 displays the syntax for building the variable in the Query Builder; copy it into the box under **Enter an expression.**

**Program 6.1: Expression Builder to Create Age Group Variable**

```
Case when AGE le 21 then "18-21 Years Old"
  when 22 le AGE le 25 then "22-25 Years Old"
  when 26 le AGE le 30 then "26-30 Years Old"
  when 31 le AGE le 40 then "31-40 Years Old"
  when 41 le AGE le 50 then "41-50 Years Old"
  when 51 le AGE then  "51+ Years Old"
end
```

Click **Next**. In the next window's field, type AGE_GROUP in the top box next to **Column Name**. Click **Finish** ▶ **Close.**

## Sum Performance Variables

The next step is to sum the performance variables CUSTOMER_ID, CUSTOMER_REVENUE, and CUSTOMER_SUBSCRIPTION_COUNT (Figure 6.5). Within the same query, on the right-hand side, put your cursor next to CUSTOMER_ID under the **Summary** column. An arrow will appear. Scroll down and select COUNT DISTINCT. For easier viewing, you can adjust the column widths with your cursor. Next, right-click the space next to CUSTOMER_REVENUE and CUSTOMER_SUBSCRIPTION_COUNT and select SUM. You'll notice that a **Summary groups** pane appears at the bottom right section of the window. The software guesses GENDER and CALCULATED AGE_GROUP, which is correct. Click **Run**.

### Figure 6.5: Sum Performance Variables across Gender and Age Group

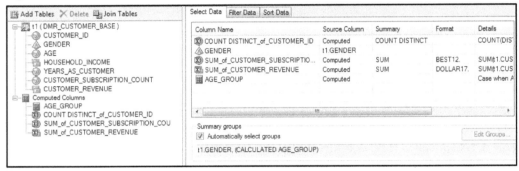

In Output 6.2, the data set appears to be summarized correctly. The new summary variables have the prefix SUM_. We will use these variables in the penetration and wallet analyses.

### Output 6.2: Summarized DMR Publishing Customer Data

| | COUNT DISTINCT_of_CUSTOMER_ID | GENDER | SUM_of_CUSTOMER_SUBSCRIPTION_COU | SUM_of_CUSTOMER_REVENUE | AGE_GROUP |
|---|---|---|---|---|---|
| 1 | 606 | F | 3448 | $64,845 | 18-21 Years Old |
| 2 | 1036 | F | 4577 | $80,295 | 22-25 Years Old |
| 3 | 1119 | F | 4166 | $78,140 | 26-30 Years Old |
| 4 | 1550 | F | 5311 | $105,465 | 31-40 Years Old |
| 5 | 1138 | F | 4188 | $80,985 | 41-50 Years Old |
| 6 | 691 | F | 2240 | $42,985 | 51+ Years Old |
| 7 | 401 | M | 2012 | $40,955 | 18-21 Years Old |
| 8 | 1239 | M | 4944 | $88,855 | 22-25 Years Old |
| 9 | 1775 | M | 6514 | $122,680 | 26-30 Years Old |
| 10 | 2805 | M | 9853 | $196,140 | 31-40 Years Old |
| 11 | 1894 | M | 6970 | $137,960 | 41-50 Years Old |
| 12 | 1258 | M | 4198 | $80,595 | 51+ Years Old |
| 13 | 50 | U | 287 | $5,310 | 18-21 Years Old |
| 14 | 122 | U | 519 | $9,315 | 22-25 Years Old |
| 15 | 94 | U | 410 | $8,480 | 26-30 Years Old |
| 16 | 115 | U | 395 | $8,105 | 31-40 Years Old |
| 17 | 77 | U | 261 | $4,910 | 41-50 Years Old |
| 18 | 49 | U | 139 | $2,620 | 51+ Years Old |

## Merge Customer and Market Data

The next step is to merge the data. If you don't see the Process Flow on the right side of your screen, double-click on **Process Flow** in the top left corner at the top of the Project Tree, and it will appear in the work area on the right as seen in Figure 6.6.

**Figure 6.6: Select Data for Merge Query**

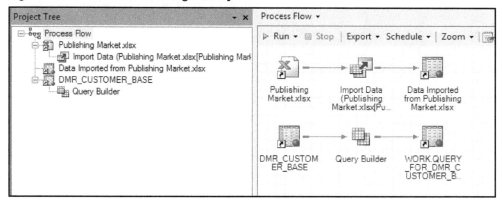

Right-click the newly summarized data and select **Query Builder**. After the window opens, drag all the variables from t1 (QUERY_FOR_DMR_CUSTOMER_BASE) into the right-hand column. In the upper left-hand corner, look for **Join tables**. A new window will open with the fields in the new data set. Click **Add Tables**. Select WORK.PUBLISHING_MARKET (Figure 6.7). If you can't see the whole name, put your cursor over the fields. The full name will appear. Click **Open**.

**Figure 6.7: Select Data for Join**

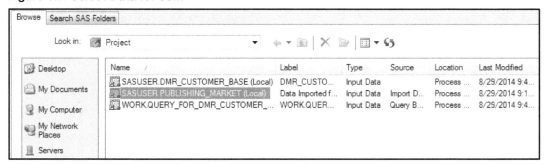

The software will guess a match variable. In this case, it guesses GENDER. But, it is an inner join as shown in the Venn diagram. Right-click on the circles and select **Properties**. You want to match the records in the DMR Publishing customer data that finds a match in the Publishing Market data; therefore, from the list at the top, select **All rows from the right table given the condition ( Right Join ).** Also, notice just below the circles that it asks if you want t2.GENDER to equal (=) t1.GENDER. The default is correct.

You are also going to match on AGE_GROUP. Right-click on AGE_GROUP in the first data set (t1). Follow the prompt to match to AGE_GROUP in the second data set (t2). Next, a window will appear, and a prompt will ask you what kind of join you want to select (Figure 6.8). As directed in the first join using GENDER, you want to match the records in the DMR Publishing customer data that finds a match in the Publishing Market data; therefore, select **All rows from the right table given the condition (Right Join )**. Also, notice just below the circles that asks if you want t2.AGE_GROUP to equal (=) t1.AGE. The default is correct. Click **Close**.

**Figure 6.8: Merge of DMR Customer and Market Data Sets**

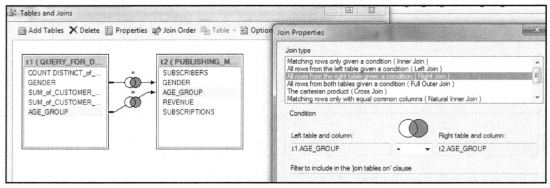

You will be back in the main screen of the query builder. From the left column, drag the following variables from the (t2) PUBLISHING_MARKET to the right side under Column Name: SUBSCRIBERS, REVENUE, SUBSCRIPTIONS (Figure 6.9).

**Figure 6.9: Add Variables from Publishing Market Data**

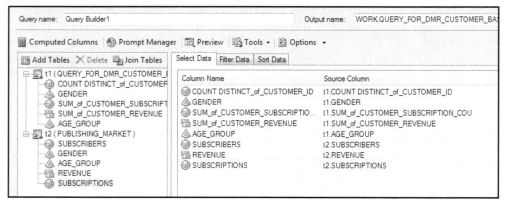

Before you run a query, you can change the column names (Figure 6.10).

**Figure 6.10: Rename Three Columns to Business-Related Names**

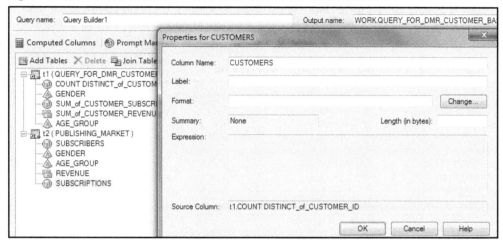

Double-click under **Column Name** on the right, where COUNT DISTINCT appears. When the window opens, type in a new name, such as CUSTOMERS. Click **OK**. Repeat the process for the two other summary columns. Suggested names are CUSTOMER_SUBS and CUSTOMER_REVENUE. Now, click **Run**.

The merged data shows a perfect match between the customer data and the market data. Notice that it is at the GENDER-by-AGE_GROUP level.

**Output 6.3: Merged Data at Gender and Age Group Level**

| | CUSTOMERS | GENDER | CUSTOMER_SUBS | CUSTOMER-REVENUE | AGE_GROUP | SUBSCRIBERS | REVENUE | SUBSCRIPTIONS |
|---|---|---|---|---|---|---|---|---|
| 1 | 606 | F | 3448 | $64,845 | 18-21 Years Old | 10,484 | $431,319 | 24,187 |
| 2 | 1036 | F | 4577 | $80,295 | 22-25 Years Old | 14,169 | $885,233 | 43,451 |
| 3 | 1119 | F | 4166 | $78,140 | 26-30 Years Old | 11,753 | $904,011 | 37,418 |
| 4 | 1550 | F | 5311 | $105,465 | 31-40 Years Old | 16,077 | $1,303,178 | 51,428 |
| 5 | 1138 | F | 4188 | $80,985 | 41-50 Years Old | 11,421 | $1,103,420 | 40,624 |
| 6 | 691 | F | 2240 | $42,985 | 51+ Years Old | 9,432 | $781,581 | 30,789 |
| 7 | 401 | M | 2012 | $40,955 | 18-21 Years Old | 8,561 | $289,773 | 16,728 |
| 8 | 1239 | M | 4944 | $88,855 | 22-25 Years Old | 15,836 | $829,967 | 42,518 |
| 9 | 1775 | M | 6514 | $122,680 | 26-30 Years Old | 16,105 | $1,091,799 | 47,651 |
| 10 | 2805 | M | 9853 | $196,140 | 31-40 Years Old | 22,319 | $1,806,043 | 69,480 |
| 11 | 1894 | M | 6970 | $137,960 | 41-50 Years Old | 15,180 | $1,458,032 | 53,122 |
| 12 | 1258 | M | 4198 | $80,595 | 51+ Years Old | 12,288 | $1,156,802 | 42,818 |
| 13 | 50 | U | 287 | $5,310 | 18-21 Years Old | 1,243 | $45,998 | 2,609 |
| 14 | 122 | U | 519 | $9,315 | 22-25 Years Old | 1,566 | $87,298 | 4,398 |
| 15 | 94 | U | 410 | $8,480 | 26-30 Years Old | 1,174 | $76,911 | 3,411 |
| 16 | 115 | U | 395 | $8,105 | 31-40 Years Old | 1,149 | $83,765 | 3,497 |
| 17 | 77 | U | 261 | $4,910 | 41-50 Years Old | 696 | $59,602 | 2,258 |
| 18 | 49 | U | 139 | $2,620 | 51+ Years Old | 571 | $42,800 | 1,728 |

For your current analysis, you want to look at penetration and share of wallet by age group. So, you will summarize the data one more time. Return to the **Process Flow** and right-click the merged data and select **Query Builder**.

When the window opens, drag each variable except GENDER to the right side and drop it under **Column Name**. Put your cursor in the empty space under **Summary**, and a dropdown menu will appear with the word NONE (Figure 6.11). Click and select SUM next to each variable except AGE_GROUP.

**Figure 6.11: Summing of Merged Data to Age Group Level**

| Column Name | Source Column | Summary |
|---|---|---|
| SUM_of_CUSTOMERS | Computed | SUM |
| SUM_of_CUSTOMER_SUBS | Computed | SUM |
| SUM_of_CUSTOMER_REVENUE | Computed | SUM |
| AGE_GROUP | t1.AGE_GROUP | |
| SUM_of_SUBSCRIBERS | Computed | SUM |
| SUM_of_REVENUE | Computed | SUM |
| SUM_of_SUBSCRIPTIONS | Computed | SUM |

When you are finished, click **Run**. This step summarizes all the customer and market values to the AGE_GROUP level. Review and close the data view.

## Calculate the Share of Wallet

The next step is to calculate the share of wallet for customers, subscriptions, and revenues with use of the query tool. With your cursor on the summarized data, right-click and then click **Query Builder**. Once it opens, pull all of the variables into the right-hand pane. You can pull each variable separately or move the entire data set at one time by dragging the icon where t1 (QUERY_FOR_DMR appears). For ease of viewing, widen the **Column Name** heading on the right side until you can see the full column names.

Next, go to **Computed Columns** ▶ **New** ▶ **Advanced Expression** ▶ **Next**. In the box under **Advanced Expression**, type SUM_of_CUSTOMERS/SUM_of_SUBSCRIBERS and click **Next** (Figure 6.12). Type PCT_of_SUBSCRIBERS next to **Column Name**. At the bottom of the window, next to **Format**, click **Change** and select **Numeric** and **Percentw.d**. Leave other settings unchanged as default. Click **OK** ▶ **Finish**. In the remaining window, click **Close**.

**Figure 6.12: Computed Columns to Create Percentage Variables**

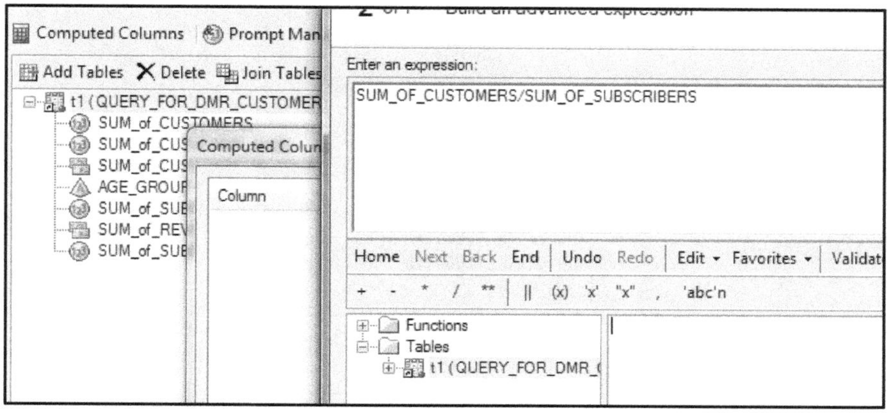

Next, click **New** and type SUM_of_CUSTOMER_SUBS/SUM_of_SUBSCRIPTIONS. Click **Next** and name the fraction PCT_of_SUBSCRIPTIONS. Create the same format and click **Finish**.

Repeat the process with SUM_of_CUSTOMER_REVENUE/SUM_of_REVENUE in the box. Use the same format. Click **Next** and name the fraction PCT_of_REVENUE. Click **Next** ▶ **Finish**. Click **Close** ▶ **Run**.

**NOTE:** To avoid errors, variable names must be exactly as stated.

Output 6.4 displays the final percentage values.

**Output 6.4: Final Percentage Variables Showing Percent of DMR Subscriber Measures as Percent of Total Market**

| | PCT_OF_SUBSCRIBERS | PCT_OF_SUBSCRIPTIONS | PCT_OF_REVENUE |
|---|---|---|---|
| 1 | 5% | 28% | 14% |
| 2 | 8% | 32% | 10% |
| 3 | 10% | 38% | 10% |
| 4 | 11% | 39% | 10% |
| 5 | 11% | 42% | 9% |
| 6 | 9% | 30% | 6% |

## Penetration and Share of Wallet

Your data is now ready to analyze. To get a clear view by AGE_GROUP, close the data and highlight the last created data set. Click **Tasks** ▶ **Describe** ▶ **List Data.** This process allows you to take the variables or columns that you select and print them in a report format. The window in Figure 6.13 allows you to manage the report output.

**Figure 6.13: Creation of a Report**

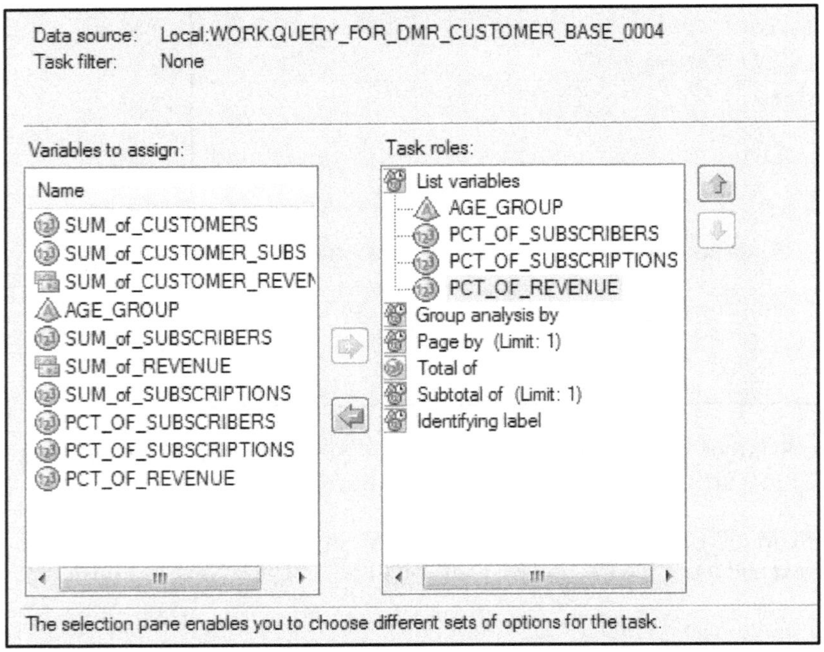

Data source:   Local:WORK.QUERY_FOR_DMR_CUSTOMER_BASE_0004
Task filter:    None

Variables to assign:

Name
- SUM_of_CUSTOMERS
- SUM_of_CUSTOMER_SUBS
- SUM_of_CUSTOMER_REVEN
- AGE_GROUP
- SUM_of_SUBSCRIBERS
- SUM_of_REVENUE
- SUM_of_SUBSCRIPTIONS
- PCT_OF_SUBSCRIBERS
- PCT_OF_SUBSCRIPTIONS
- PCT_OF_REVENUE

Task roles:
- List variables
  - AGE_GROUP
  - PCT_OF_SUBSCRIBERS
  - PCT_OF_SUBSCRIPTIONS
  - PCT_OF_REVENUE
- Group analysis by
- Page by  (Limit: 1)
- Total of
- Subtotal of  (Limit: 1)
- Identifying label

The selection pane enables you to choose different sets of options for the task.

Drag AGE_GROUP and the three percentage (PCT) variables to the right side. Click **Run**. Look for the output created in Microsoft Word.

## Results

Reporting your results is one of the most important steps in your analysis. If you do everything right and can't communicate the results, your efforts and insights will be wasted.

In Output 6.5, the final report displays the market penetration analysis and share of wallet by age group. When you are evaluating the percentage of subscribers, the lowest market penetration is in the 18 to 21-year-old age group. It is higher with 26 to 50-year-olds, at 10% to 11%. It drops slightly at age 51 or older (9%).

**Output 6.5: Final Penetration Analysis by Age Group**

| Row number | AGE_GROUP | PCT_OF_SUBSCRIBERS | PCT_OF_SUBSCRIPTIONS | PCT_OF_REVENUE |
|---|---|---|---|---|
| 1 | 18-21 Years Old | 5% | 28% | 14% |
| 2 | 22-25 Years Old | 8% | 32% | 10% |
| 3 | 26-30 Years Old | 10% | 38% | 10% |
| 4 | 31-40 Years Old | 11% | 39% | 10% |
| 5 | 41-50 Years Old | 11% | 42% | 9% |
| 6 | 51+ Years Old | 9% | 30% | 6% |

The percentage of subscriptions shows a different outcome. The youth, aged 18 to 21 years, have a higher number of subscriptions per person. So, they bring in the largest percentage of the revenue. Their share of wallet is 14%, compared with only 6% for those aged 51 or older.

Several conclusions can be drawn from this analysis. If you are looking for additional subscribers that bring in high average revenue, the 18- to 21-year-old market is the place to focus. These subscribers seem to be happiest with the current offerings, although there is still much room to grow.

On the other hand, growth opportunities for the business may emerge in the other age groups if new products are emphasized. The share of wallet is 10% or less for customers aged 22 years or older. This result suggests that products could be developed to appeal to groups older than 21.

## Notes from the Field

This case study provides some rich insights and potential opportunities for DMR Publishing. What the company does with the information depends much more on the abilities of the organization.

Consider that there are two new opportunities identified. The first opportunity is to find additional subscribers that look like their current subscriber base. The job of finding additional subscribers might be handled by the marketing department. The second opportunity is to create new products for the underserved areas of the market. This need for new products would likely be a project for the product development side of the business. The optimal solution may be a mixture of both departments.

Because both market growth for the current business and product development are important, management must make the decision about how to allocate resources. The best outcomes are seen when companies have fostered a culture of big data so that departments are able to collaborate and partner in the implementation of the analysis results, optimizing profits for the overall company.

# Chapter 7: Cluster Analysis Using SAS Enterprise Miner

## Introduction

In Chapter 5, you explored the DMR Publishing customer data to familiarize yourself with the data and look for customer patterns and trends. In Chapter 6, you compared the performance of DMR Publishing customers to the performance of the entire publishing market. In this chapter, your goal is to gain a deeper understanding of your customer base by grouping your customers into segments that have similar characteristics. By unveiling the similarities and differences among your customers, you can design marketing programs and database enrichment strategies that align with your long-term strategic goals.

## Project Overview

The leadership team at DMR Publishing Company has asked you to help it gain a better understanding of its customer base. The team wants to know whether all their customers look alike, or whether they naturally segment into different groups. The segmentation is performed with SAS Enterprise Miner.

The project has eight steps:

1. Initiate the project in SAS Enterprise Miner 13.1.
2. Input data and assign variable roles.

3. View variable distributions and transform if necessary.
4. Filter data.
5. Build clusters.
6. Build segment profiles.
7. Recommend business actions.

## Cluster Analysis

*Cluster analysis*, or *clustering*, is a process that places observations into groups or segments that favor similarity within each group while favoring dissimilarity between groups. This ability to group or segment customers can be useful if you want to market to a group of your customers who look alike. You may have information about your customers, but you don't know what makes them similar or different. Cluster analysis performs this type of grouping.

The clustering methods in the SAS Enterprise Miner cluster node perform disjoint cluster analysis by calculating the Euclidean distances between at least one quantitative variable and seeds. The *seeds* are the original centers of the clusters. The centers of the clusters change during the clustering process. You can control the clustering criterion that is used to measure the distance between data observations and seeds. The final clusters are mutually exclusive in that no observation populates more than one cluster. This feature of the cluster node makes it very useful for business purposes.

### Initiate the Project

To open SAS Enterprise Miner, click the icon on your desktop or Start menu. Your first choice is to open an existing project or create a new project. Highlight and click **New Project** (Figure 7.1).

> **NOTE:** It is possible to do basic cluster analysis in SAS Enterprise Guide. But SAS Enterprise Miner has automated and streamlined the process that is optimal for creating mutually exclusive clusters. This ability to create mutually exclusive clusters is essential for use in marketing and risk analysis.

**Figure 7.1: Initialize SAS Enterprise Miner**

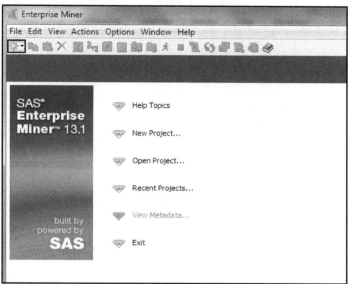

A window will open that asks you to name your project and select a SAS Server Directory (Figure 7.2). Depending on your setup, additional connections may be required. If this is the case or if you are not sure how to locate your data, contact your information technology department or other technical assistant. Otherwise, click browse and select a folder in which you would like to save your project files.

**Figure 7.2: Create and Name New Project**

When you are finished, click **Next** ▶ and you will see window that summarizes your options. Click **Finish**. You are now in the EM workspace, as shown in Figure 7.3.

**Figure 7.3: View SAS Enterprise Miner Workspace**

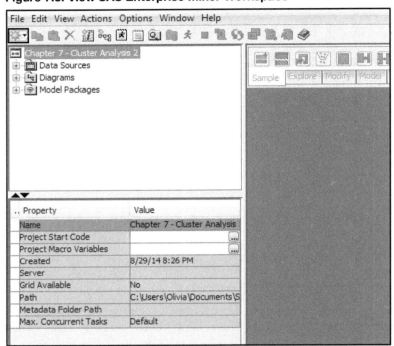

## Input the Data Source and Assign Variable Roles

Next, double click on the Data Sources icon (upper left-hand menu, directly under the word ***Actions***). The Data Import Wizard will open, asking you to create a SAS Table (Figure 7.4).

**Figure 7.4: Locate Data Source**

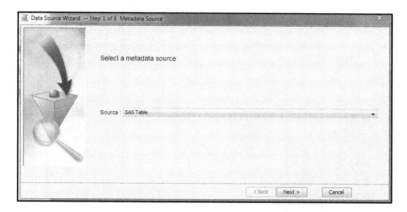

Click **Next** ▶and browse for the DMR_CUSTOMER_BASE data set created in Chapter 3. Once you locate the data, select the data and click **OK**, then click **Next** ▶. The next window shows you a summary. Click **Next** ▶. For Meta Data Advisory options, click **Advanced** and **Next** ▶. A window appears that displays the variable characteristics and offers options for exploring the distributions (Figure 7.5).

**Figure 7.5: Assign Variable Roles and Explore Variables**

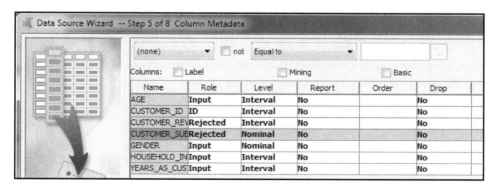

First, you need to change the role of CUSTOMER_REVENUE and CUSTOMER_SUBSCRIPTION_COUNT to "Rejected." These are outcome variables that will be used in future chapters. But for now, you do not want to include them in your analysis.

**NOTE: It is necessary to have an ID variable for clustering in order to track the observations in each cluster.**

Highlight AGE and click **Explore**. A large window opens and has four quadrants. Maximize the quadrant in the lower left to get Output 7.1.

**Output 7.1: View Distribution of the Age Variable**

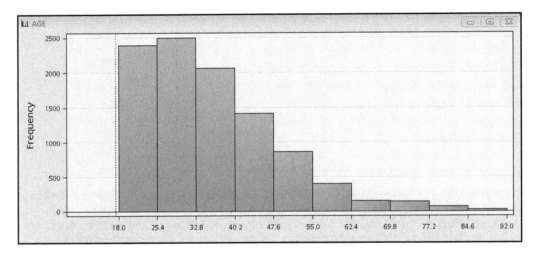

For clustering, you want variables to have a bell-shaped curve that represents a normal distribution. Because AGE is not normally distributed, you can use the **transform** function. Because AGE is skewed to the right, or positively skewed, you should use a log transformation after all the variables are explored.

**NOTE:** To close the **Results** window in SAS Enterprise Miner, click the **X** in the upper right-hand corner. It will close only the results window; the main project will remain open.

Explore the two remaining continuous variables, HOUSEHOLD_INCOME and YEARS_AS CUSTOMER, by following the same process. Because both variables are skewed to the right, these variables will also need to be transformed with use of the log transformation. But first, you must complete the import process. Click **Next ►** several times and then **Finish.**

Once the DMR_CUSTOMER_BASE data source is created, the data set name will appear under Data Sources in the upper diagram as shown in Figure 7.6.

**Figure 7.6: Open Project View with Data Source in Clustering Diagram**

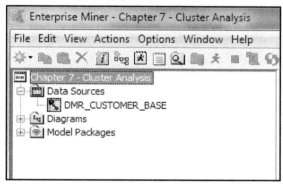

Next, you want to create a workspace for your cluster analysis. Go to the top menu and click **File ► New ► Diagram**. When the box opens, name the diagram Clustering and click **OK**. A work area will appear on the right. Place your cursor on the data set, DMR_CUSTOMER_BASE, and drag it into the clustering work area on the right.

## Transform Variables

Next, look at the menu above the diagram and click the **Modify** tab. The last icon in the row above is **Transform Variables**. Drag the **Transform Variables** icon to the **Diagram** and connect it to **DMR _CUSTOMER** data with an arrow (Figure 7.7).

**Figure 7.7: Use the Transform Variable Icon**

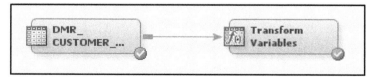

To connect the DMR Customer data to the Transform Variables node, right-click on the **Transform Variables** node and select **Run.** When offered to view the results, click **OK.**

With the Transform Variables node still highlighted, look to the lower left of the work area shown in Figure 7.8.

**Figure 7.8:  Locate the Transform Variable Formula Menu**

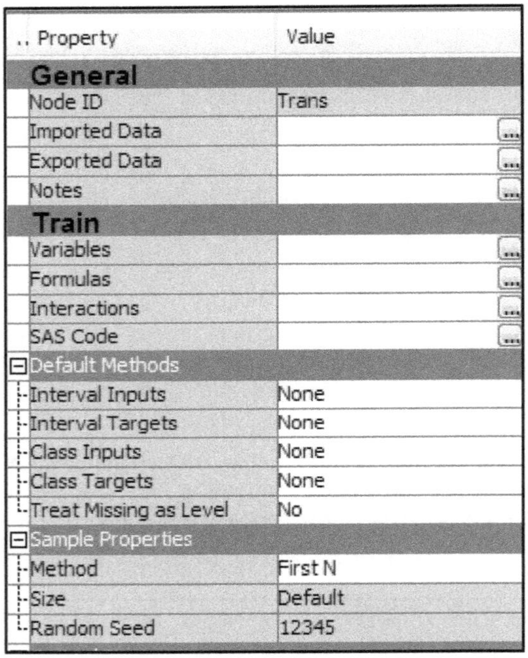

| .. Property | Value |
|---|---|
| **General** | |
| Node ID | Trans |
| Imported Data | ... |
| Exported Data | ... |
| Notes | ... |
| **Train** | |
| Variables | ... |
| Formulas | ... |
| Interactions | ... |
| SAS Code | ... |
| ⊟Default Methods | |
| ┊-Interval Inputs | None |
| ┊-Interval Targets | None |
| ┊-Class Inputs | None |
| ┊-Class Targets | None |
| └-Treat Missing as Level | No |
| ⊟Sample Properties | |
| ┊-Method | First N |
| ┊-Size | Default |
| └-Random Seed | 12345 |

Under **Train**, next to **Formulas**, click the three dots to the right. A window will open that shows all the variables and each distribution, depending on which variable is selected (Figure 7.9).

**Figure 7.9: View the Transformation Overview**

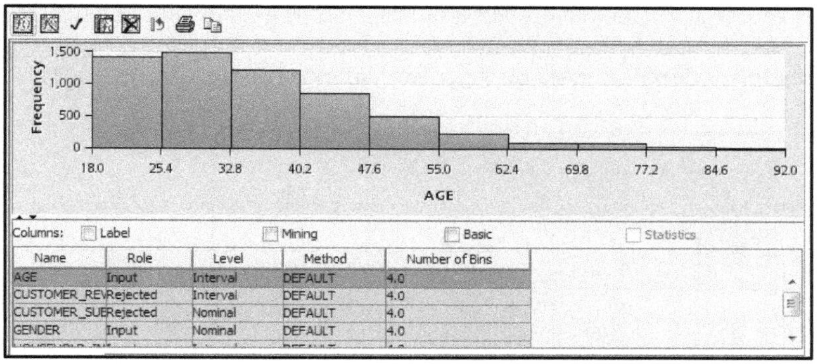

Highlight the variable AGE. In the upper left-hand corner, click the **Create** icon. The word 'Create' will appear when you hover over the icon. The box in Figure 7.10 will appear.

**Figure 7.10: Transform the Age Variable**

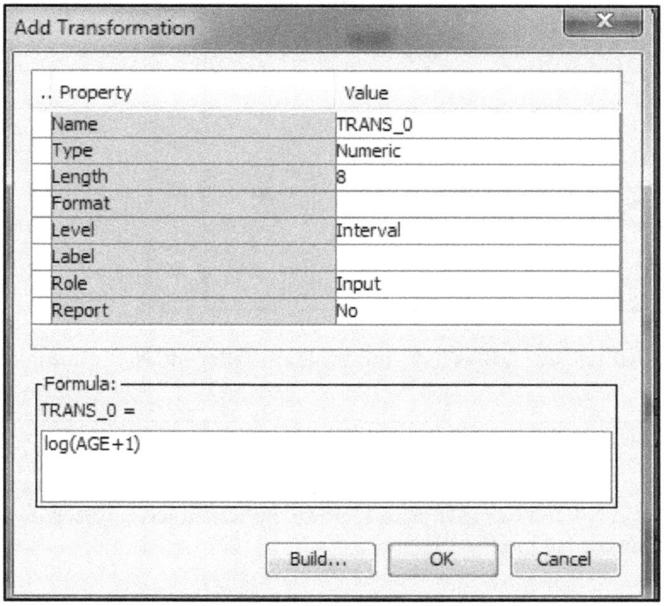

To the right of **Property** under **Value**, change the name from **TRANS_01** to **AGE_LOG**. Repeat the process for HOUSEHOLD_INCOME and YEARS_AS_CUSTOMER, using similar **Names** and log transformations based on the formulas in Figure 7.10. After each formula is typed in, click **OK**. Note that the log of HOUSEHOLD_INCOME becomes HH_INCOME_LOG and equals log (HOUSEHOLD_INCOME+100). The name for log of YEARS_AS_CUSTOMER is YEARS_AS_CUST_LOG.

The transformation formulas appear at the bottom of the Formulas window as shown in Figure 7.11.

**Figure 7.11: Create the Transformation Formulas for Age, Household Income, and Years as Customer**

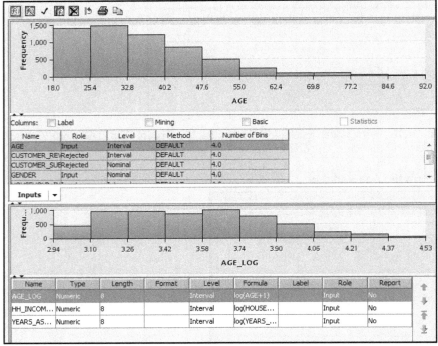

Click **OK**. Then, right-click on the **Transform Variables** icon and hit **Run**. After the run is complete, click **OK**.

## Filter Data

Because clustering is sensitive to outliers, you will get better results if you run your variable through a filtering process. Above your diagram, click on the **Sample** tab and go to the fourth icon, **Filter**. Drag it onto the diagram and connect it with an arrow (Figure 7.12).

**Figure 7.12:  Filter the Data**

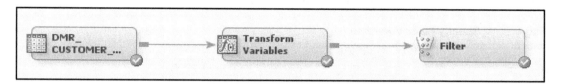

Right click on the **Filter** node and select **Run**.

To look for potential outliers, highlight the filter icon, go to the lower left menu, and click the three dots to the right of **Interval Variables** (Figure 7.13).

**Figure 7.13: Locate the Filter Menu**

| .. Property | Value |
|---|---|
| **General** | |
| Node ID | Filter |
| Imported Data | |
| Exported Data | |
| Notes | |
| **Train** | |
| Export Table | Filtered |
| Tables to Filter | Training Data |
| Distribution Data Sets | Yes |
| ⊟ Class Variables | |
| ┊·Class Variables | |
| ┊·Default Filtering Method | Rare Values (Percentage) |
| ┊·Keep Missing Values | Yes |
| ┊·Normalized Values | Yes |
| ┊·Minimum Frequency Cutoff | 1 |
| ┊·Minimum Cutoff for Percent | 0.01 |
| ┊·Maximum Number of Levels | 25 |
| ⊟ Interval Variables | |
| ┊·Interval Variables | |
| ┊·Default Filtering Method | Standard Deviations from t |
| ┊·Keep Missing Values | Yes |

This dropdown option opens a window that allows you to view each distribution (Figure 7.14). For HH_INCOME_LOG, set the **Filter Lower Limit** to 8 and click **Apply Filter**.

**Figure 7.14: Specify Filter Variables**

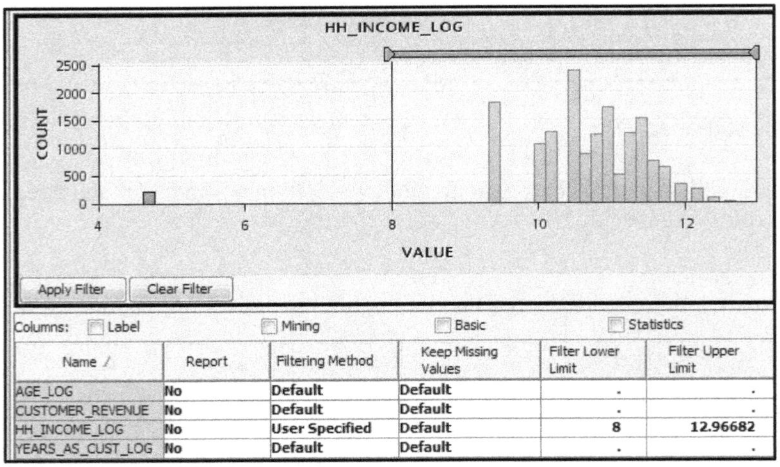
HH_INCOME_LOG
COUNT
2500
2000
1500
1000
500
0
4
6
8
10
12
VALUE
Apply Filter
Clear Filter
Columns: Label
Mining
Basic
Statistics
Name
Report
Filtering Method
Keep Missing Values
Filter Lower Limit
Filter Upper Limit
AGE_LOG No Default Default . .
CUSTOMER_REVENUE No Default Default . .
HH_INCOME_LOG No User Specified Default 8 12.96682
YEARS_AS_CUST_LOG No Default Default . .

Click **OK.** Then right-click the **Filter** icon and select **Run**. When it completes, click **OK**.

## Build Clusters

Above the diagram window, go to the **Explore** tab and pick the second icon, **Cluster**. Drag the **Cluster** icon onto the diagram and connect it with the **Filter** node (Figure 7.15).

**Figure 7.15: Add the Cluster Process**

DMR_
CUSTOMER_...
Transform
Variables
Filter
Cluster

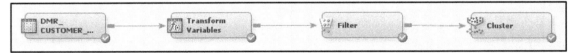

When you highlight the Cluster icon, the Property menu appears on the lower left (Figure 7.16).

**Figure 7.16: View the Property Menu for the Add-Cluster Process**

| Property | Value |
|---|---|
| **General** | |
| Node ID | Clus |
| Imported Data | |
| Exported Data | |
| Notes | |
| **Train** | |
| Variables | |
| Cluster Variable Role | Segment |
| Internal Standardization | Range |
| Number of Clusters | |
| Specification Method | Automatic |
| Maximum Number of Cluster | 10 |
| Selection Criterion | |
| Clustering Method | Centroid |
| Preliminary Maximum | 50 |
| Minimum | 2 |
| Final Maximum | 20 |
| CCC Cutoff | 3 |
| Encoding of Class Variables | |
| Ordinal Encoding | Rank |
| Nominal Encoding | GLM |
| Initial Cluster Seeds | |
| Seed Initialization Method | Full Replacement |
| Minimum Radius | 0.0 |
| Drift During Training | No |

Continuous variables come in different scales, such as counts, minutes, and dollars. You will want to standardize these variables for clustering. Otherwise, the variables with the higher scale will have an advantage. Under **Train**, change **Internal Standardization** to **Range**. This option standardizes the values for each variable to a value between 0 and 1. For **Clustering Method**, select **Centroid**. This method is usually better than the Ward method for handling contrasting data. Change the **Seed Initialization Method** to **Full Replacement** to select seeds that are well-separated. Right click, then click **Run**. When the run is complete, click on **Results**.

Output 7.2 shows the upper left and lower left quadrants of in the Results windows four quadrants. Each quadrant offers insights into the results of the process as defined:

- *Segment Plot*—the upper left quadrant displays a segment plot of the clustering variables with the highest importance. The results show how the values of age, log, and gender are distributed among the clusters. To see the value of the segment and other statistics, place your cursor on the color within each segment.

- *Segment Size*—the lower left quadrant offers a visual display of the size of each cluster in a pie chart. To view the number of customers in each cluster, hover your cursor over each segment of the pie chart.

Because the next step is to build segment profiles, the main interest in these four quadrants is the number of clusters displayed in the lower left quadrant. The pie chart shows that there are five total clusters. Four to eight clusters is a good amount when building segment profiles.

**Output 7.2: View the Cluster Results**

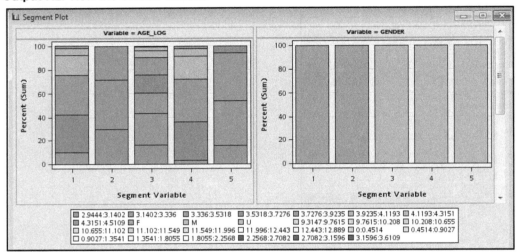

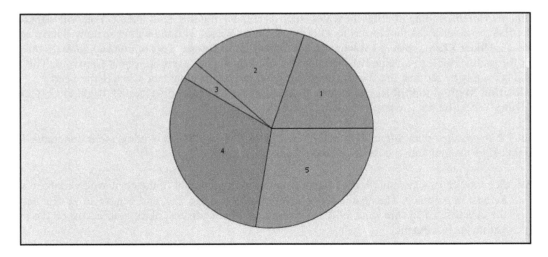

When you are finished viewing the cluster results, close the Results window and return to the diagram.

## Build Segment Profiles

In the menu above the diagram, click **Assess** and **Segment Profile**. Drag the Segment Profile icon into the diagram and connect with an arrow (Figure 7.17).

**Figure 7.17: Add the Segment Profile Process**

Right-click and **Run**. When the results window appears, click **Results**.

The initial Results window displays four quadrants. Focus on the upper right quadrant, which displays the profiles of each segment. Maximize the upper right quadrant for a closer look, as partially seen in Output 7.3.

**Output 7.3: View the Segment Profile Results**

Maximize the upper right quadrant for a closer look, as seen in Output 7.4.

This output displays the characteristics of the customers in each segment. The segments are arranged in order of the size of the segment. The segment with the greatest number of customers is listed first. The segment with the least number of customers is listed last:

Segment 4 has 4,885 customers. AGE_LOG is the strongest predictor. The solid bars represent the distribution within the segment. The dark red outline represents the overall population. So this cluster contains customers that are older than average. They have also been customers longer than average. The third characteristic is gender. The customers in this cluster are all male. If you want to see the values, right-click next to the circle and select **Expand**. Then, hover over the edge of the outside circle, and a window will appear that gives you values for each area. Finally, the household income log shows the same trend toward higher than average.

Segment 5 has 4,354 customers. AGE_LOG is the strongest predictor. This cluster contains customers that are younger than average. They have also been customers for less time than average. The customers in this cluster are all male. Finally, household income shows the same trend towards lower than average.

Segment 1 has 3,111 customers. GENDER is the strongest predictor. This cluster is all female. AGE is next. This cluster contains customers that are older than average. They have also been customers longer than average. Finally, household income shows the same trend towards slightly higher than average.

Segment 2 has 2,941 customers. GENDER is the strongest predictor. This cluster is all female. This cluster contains customers that are younger than average. They have also been customers less time than average. Finally, household income shows the same trend toward lower than average.

Segment 3 has only 499 customers. Its only characteristic is GENDER, which is all unknown. This result doesn't show in Output 7.4, but is visible when the process is run in SAS Enterprise Miner.

Another useful result is the **Variable Worth** in the lower left quadrant of the **Results** window, as shown in Output 7.4.

**Output 7.4: View Variable Worth**

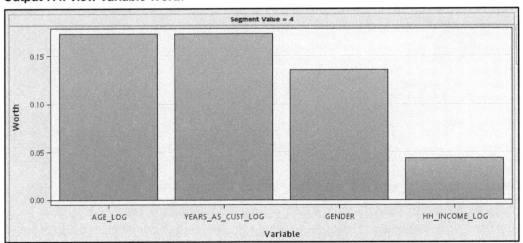

AGE_LOG is the strongest contributor to the cluster segments, followed by YEARS_AS_CUST_LOG, which is very close. GENDER is third in importance as a contributor. And, finally, HH_INCOME_LOG is the least powerful contributor to the cluster segments.

## Analyze Clusters and Recommend Marketing or Product Development Actions

This set of results informs DMR Publishing that it can focus its marketing and product development for different age groups with consideration for loyalty (years as customer), gender, and household income.

Because Segment 4 is the largest, it is a good place to focus your analysis. However, one thing to notice is the similarity between Segment 4 and the third largest segment, Segment 1. They both have loyal customers (based on years as customer) who are older than average and have slightly higher than average income. The main difference is their gender. Together, Segment 4 and Segment 1 represent 50% of the DMR Publishing customer base. Therefore, you may want to consider a two-level approach:

Consider separate marketing actions or product development for gender-specific publications, such as men's health or women's fashion magazines. Or, your approach might be as simple as different magazine covers or advertisements aimed at each gender.

Consider combining these two segments for publications that are not gender-specific, such as cooking and travel magazines or business journals.

Segment 5, the second largest cluster, offers another opportunity. This group of all male, younger-than-average customers is a prime audience for publications that appeal to that demographic group. If DMR Publishing doesn't already offer some sports or technology magazines, you might suggest that they add some to their list of publications.

## Notes from the Field

Segment profiles are used to describe the clusters and make them actionable. As you have seen, the patterns revealed and the insights that emerge can guide creative marketing decisions and well as product or service development. Once you understand your customers based on your existing data, consider purchasing additional data to enrich your clusters, enhance your analysis, and grow your customer base. As discussed in Chapter 2, there are many good sources of external data. You can purchase characteristics such as hobbies and interests, buying patterns, and social or online behavior and append them to your existing customer base. Once your customer data is enriched, you can rerun your cluster segments and refine your marketing and product development strategies.

When sharing your results or proposing strategic initiatives with your stakeholders and end-users, speak in terms of their business objectives and relate your recommendations to the strategic goals of the company.

# Chapter 8: Tree Analysis Using SAS Enterprise Miner

## Introduction

In several earlier chapters, we focused on understanding the customers at DMR Publishing. Understanding your customers is always a good first step in any comprehensive analysis. In this chapter, you will learn how to discover and measure patterns to predict future behavior from a combination of past behaviors and characteristics. The main technique you will use in this process is tree analysis. One of the benefits of tree analysis is that in addition to being a powerful predictive technique, it also has a strong descriptive component. The descriptive results of a tree analysis are easily interpreted, making this technique a favorite among business analysts and marketers.

## Project Overview

The leadership team at DMR Publishing Company is interested in understanding the characteristics of customers that subscribe to more than one magazine or journal when compared to customers that have only one subscription. For this analysis, you will use SAS Enterprise Miner.

This project has ten steps:

1. Initiate the project in SAS Enterprise Miner 13.1.
2. Input the data source.
3. Define the variable roles.
4. Define the target variable.
5. Partition the data.
6. Set the tree properties.
7. Build the decision tree.
8. Adjust your graph properties.
9. Interpret your findings.
10. Display Node Rules.

# Decision Tree Analysis

The decision tree technique is designed to segment data according to series of simple rules. For a full documentation of the process, access the **Help** section of SAS Enterprise Miner. Once you have a project open, go to **Help ▶ Contents.** When the window opens, look for **Node Reference ▶ Model ▶ Decision Tree Node.**

## Initiate the Project

To open SAS Enterprise Miner (EM), click on the node on your desktop or Start menu. Your first choice is to open an existing project or create a new project. Highlight and click **New Project** (Figure 8.1).

**Figure 8.1: Initialize SAS Enterprise Miner**

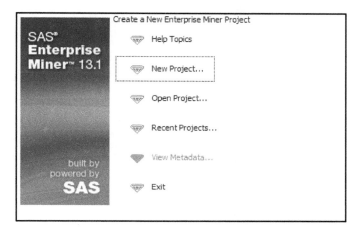

A window opens that asks you to name your project and select a SAS Server Directory (Figure 8.2). Depending on your setup, it may require additional connections. If that is the case, contact your IT department. Otherwise, click browse and select a folder where you would like to save your project files.

**Figure 8.2: Create and Name New Project**

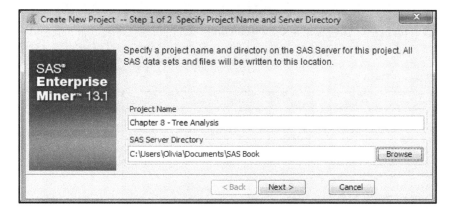

When you are finished, click **Next ▶** and you will see window that displays your choices. Click **Finish**. You are now in the EM workspace, as shown in Figure 8.3.

114 *Business Analytics Using SAS Enterprise Guide and SAS Enterprise Miner*

**Figure 8.3: SAS Enterprise Miner Workspace**

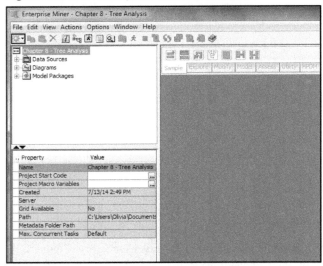

## Input the Data Source

Next, click on the **Create Data Source** node (upper left menu right under the word 'Actions.'). The Data Source Wizard will open and ask you to locate a SAS Table. Click **Next ▶** and browse in the SASUSER library for the **DMR_CUSTOMER_DATA** data set. Once you locate the data, select the data and click **OK** (Figure 8.4).

**Figure 8.4: Locate and Input Data Source**

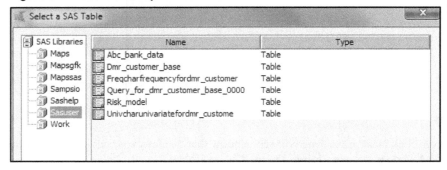

Click **Next**. The next window shows you a summary. Click **Next** again and select **Advanced** to open a table that will allow you to change various aspects of the data.

Change the **Role** for CUSTOMER_REVENUE, CUSTOMER_SUBSCRIPTION_COUNT, and YEARS_AS_CUSTOMER to **Rejected** (Figure 8.5). The role for these three variables is set to **Rejected** because they are a measure of performance with DMR Publishing. So, the tree will be built using AGE, GENDER, and HOUSEHOLD_INCOME.

Click **Next** several times until you see **Finish**. Click **Finish.** The target variable will be built using the CUSTOMER_SUBSCRIPTION_COUNT variable in a later step.

**Figure 8.5: Assign Variable Roles**

| Name | Role | Level | Report | Order | Drop |
|---|---|---|---|---|---|
| AGE | Input | Interval | No | | No |
| CUSTOMER_ID | ID | Interval | No | | No |
| CUSTOMER_REVENUE | Rejected | Interval | No | | No |
| CUSTOMER_SUBSCRIPTION_COUNT | Rejected | Nominal | No | | No |
| GENDER | Input | Nominal | No | | No |
| HOUSEHOLD_INCOME | Input | Interval | No | | No |
| YEARS_AS_CUSTOMER | Rejected | Interval | No | | No |

The **DMR_CUSTOMER_DATA** data now appears under **Data Sources** in your Project Tree display in the upper left corner of your workspace.

Next, you want to create a new diagram in your workspace. Go to **File ▶ New ▶ Diagram**. Name the diagram 'Tree Analysis.' Then, drag the DMR_CUSTOMER_DATA data into the new diagram. Right-click on the DMR_CUSTOMER_DATA icon and select **Update.** When it completes, click **OK.**

## Create Target Variable

You are now ready to create the target variable. Click the **Modify** tab and select drag the **Transform Variables** icon into the diagram. Connect the DMR_CUSTOMER_DATA icon to the **Transform Variables** icon (Figure 8.6). Right-click the **Transform Variables** icon and select **Update.** When it completes, click **OK.**

**Figure 8.6: Connect Transform Variables to DMR Customer Data**

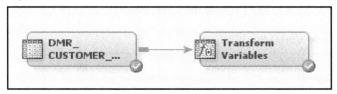

With the **Transform Variables** icon highlighted, go to the **Property** window in the lower left of the screen. Under **Train**, double-click on the three dots to the right of the **Formulas** to open a window that will allow you to create new variables (Figure 8.7). Highlight the variable CUSTOMER_SUBSCRIPTION_COUNTS. To create the target variable, click the Create icon in the upper left corner. This icon will open the window shown in the smaller window in Figure 8.7.

**Figure 8.7: Build Target Variable Equation**

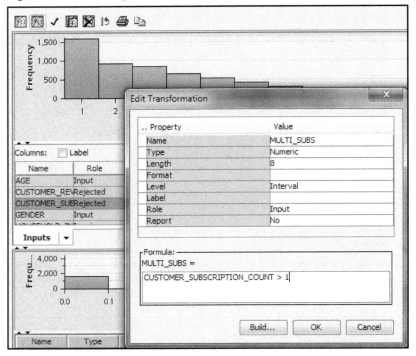

In the bottom window under Formula, type CUSTOMER_SUBSCRIPTION_COUNT > 1. In the upper portion of the smaller box to the right of Name and under Value, type MULTI_SUBS as the name of the target variable. Click **OK ▶ OK**. The name of the new target variable is now MULTI_SUBS.

Your next step is to assign MULTI_SUBS as a target variable. Click on the **Utility** tab, drag the **Metadata** icon onto the diagram, and connect it to the **Transform Variables** icon (Figure 8.8).

**Figure 8.8: Attach Metadata to Transform Variables**

Right-click on the **Metadata** icon and select **Update.** When it completes, click **OK.**

Next, highlight the **Metadata** icon and go to the **Property** window in the lower left corner. Click on the three dots under **Variables** to the right of **Train**. The window shown in Figure 8.9 opens. Under **New Role**, change the role for MULTI_SUBS to Target. Click **OK**. Right-click on the **Metadata** icon and select **Run**. When it completes, click **OK**.

**Figure 8.9: Assign Target Variable Role Using Metadata**

| Name | Hidden | Hide | Role | New Role |
|---|---|---|---|---|
| AGE | N | Default | Input | Default |
| CUSTOMER_ID | N | Default | ID | Default |
| CUSTOMER_REV | N | Default | Rejected | Default |
| CUSTOMER_SUE | Y | Default | Rejected | Default |
| GENDER | N | Default | Input | Default |
| HOUSEHOLD_IN | N | Default | Input | Default |
| MULTI_SUBS | N | Default | Input | Target |
| YEARS_AS_CUS | N | Default | Rejected | Default |

Your next step is to partition the data.

## Partition the Data

The purpose of partitioning the data is to allow you to develop the model on one subset of the data and validate the model on another subset of the data. The subsets of the data are mutually exclusive in that they do not share any common observations.

In decision tree modeling, the subsets are used as follows:

- *Train:* The train subset for the initial model fitting.
- *Validation:* The validation data set is used to evaluate the model during the tree building process and create the best subtree.

From the **Sample** tab, select the second icon from the left, **Data Partition,** and drag it onto the diagram. Connect it to the **Metadata** icon (Figure 8.10).

**Figure 8.10: Attach Data Partition to Metadata Icon**

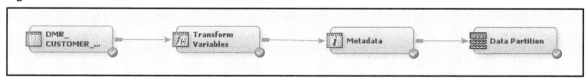

Within the **Property** window in the bottom left, look for **Data Set Allocations** under **Train.** The default shows 40% of the file allocated to **Training** and 30% each to **Validation** and **Testing.** For tree models, you only need **Training** and **Validation.** So, set **Training** and **Validation** to 50% each, and **Test** to 0% (Figure 8.11).

**Figure 8.11: Reassign Data Allocation Percentages**

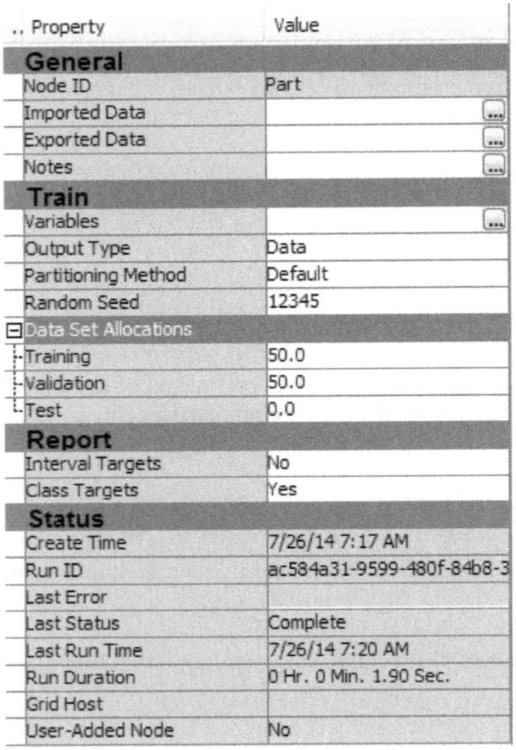

| .. Property | Value |
|---|---|
| **General** | |
| Node ID | Part |
| Imported Data | |
| Exported Data | |
| Notes | |
| **Train** | |
| Variables | |
| Output Type | Data |
| Partitioning Method | Default |
| Random Seed | 12345 |
| ⊟ Data Set Allocations | |
| ├ Training | 50.0 |
| ├ Validation | 50.0 |
| └ Test | 0.0 |
| **Report** | |
| Interval Targets | No |
| Class Targets | Yes |
| **Status** | |
| Create Time | 7/26/14 7:17 AM |
| Run ID | ac584a31-9599-480f-84b8-3 |
| Last Error | |
| Last Status | Complete |
| Last Run Time | 7/26/14 7:20 AM |
| Run Duration | 0 Hr. 0 Min. 1.90 Sec. |
| Grid Host | |
| User-Added Node | No |

Right-click on the **Data Partition** icon and select **Update.** When it completes, click **OK.** You are now ready to build your tree.

## Build the Decision Tree

In the **Model** menu above the diagram, drag the **Decision Tree** icon (second from left) into the workspace, and connect it to the **Data Partition** icon (Figure 8.12).

**Figure 8.12: Connect Decision Tree Icon to Metadata Icon**

To update the path to the decision tree node, right-click on it and select **Update.**

Highlight the **Decision Tree** node and go to the lower left menu (Figure 8.13). The defaults are good, but you might want to consider several settings. For each, click the three dots to the right of your choice:

- *Variables:* This setting is a chance to suppress variables from the decision tree. To do so, go to the **Use** column and change the value to **No**. This setting is not necessary for your current tree, but it is a way to explore other options.
- *Interactive*: This setting enables you to hand-build your tree by allowing you to create splits and prune your tree on the fly. It is beyond the scope of this book. But once you've mastered the basic steps, you might want to explore some of these options.
- *Maximum Branch:* Although this setting remains at 2 for this demonstration, in practice, one might typically use 4 or 5.
- *Maximum Depth:* Normally, you might leave this setting at 6; it is set to 3 for this demonstration.

**Figure 8.13: Assign Decision Tree Settings**

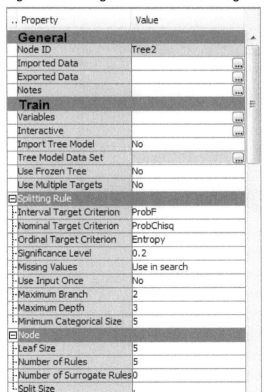

Right-click the decision tree node and click **Run** and **Yes**. When the **Run Completed** box opens, click **Results**.

## View the Decision Tree Output

The first window contains six panels. Maximize the panel in the upper right corner to display the tree shown in Outputs 8.1 and 8.2. The tree is clearly visible and you can view it in its entirety by scrolling to the left and right.

**Output 8.1: View Initial Tree (Left Branch)**

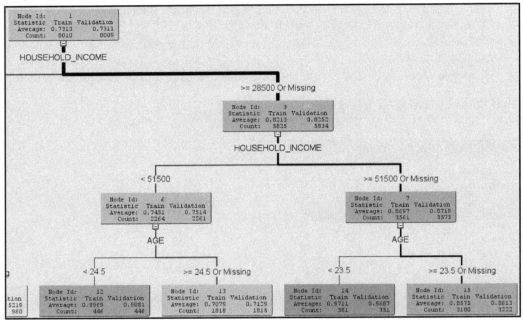

**Output 8.2: View Initial Tree (Right Branch)**

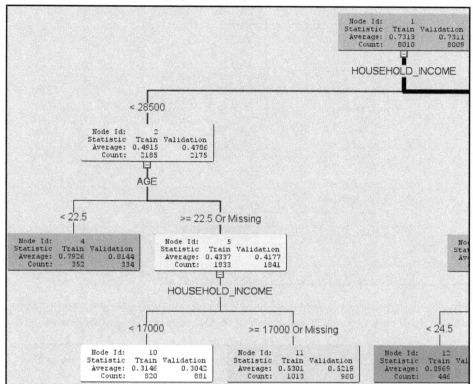

## Graph Properties

Before you begin analyzing the results, you have some options for changing the view in accordance with your personal preferences. To explore the many options within the tree output, right-click anywhere on the screen shown in Output 8.1 and select **Graph Properties**. When the window opens, click **Tree**. Change the **Orientation** to **Horizontal,** and click **OK** (Figure 8.14).

**Figure 8.14: Change Graph Properties for Decision Tree Output**

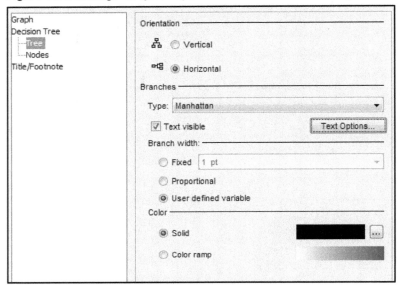

You can also click **Text Options** and change the font to **Arial Narrow** and the size to **10 pt**. Doing so is useful if the variable names are long. Click **Apply ▶ OK ▶ OK**. The tree is now in a horizontal view.

The horizontal view of the tree is shown in Output 8.3. The tree results read from left to right. Each rectangular box is called a *node*. Each node location contains information about the group of DMR customers within that node.

**Output 8.3: View Horizontal Tree**

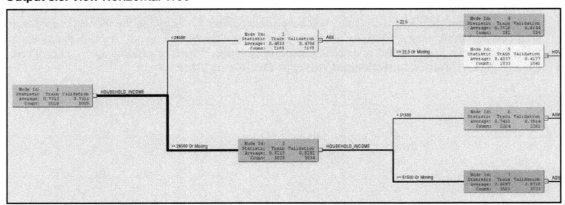

## Diagnostics

With all predictive models, there is a risk of over-fitting. In other words, your model could fit your data so well that it does not work well on new data when you implement the model. You can check for over-fitting by viewing the Subtree Sequence Plot. To view the Subtree Sequence Plot, go to the upper left menu and select **View ▶ Model ▶ Subtree Assessment Plot** (Figure 8.15).

**Figure 8.15: Access Subtree Assessment Plot**

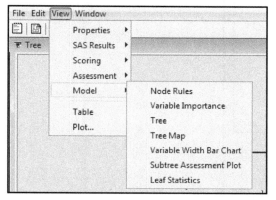

The Subtree Assessment Plot shows that the Average Squared Error continues to decrease as the number of leaves increases (Output 8.4).

**Output 8.4: View Subtree Assessment Plot**

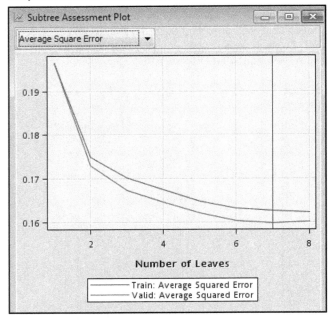

If the error begins to increase, you should limit the number of leaves or prune the tree. For more information on pruning decision trees, see *Decision Trees for Analytics Using SAS® Enterprise Miner*, by Barry de Ville and Padraic Neville (2013).

## Node Characteristics

Three dimensions of the tree output help you visually interpret the results:

- *Shading:* When the shading is darker, the percentage of the target group MULTI_SUBS > 1 is higher.
- *Line Width:* The width or thickness of the connecting lines represents the volume of records going to the node.
- *Node Values:* Each node displays values for the records for both the model and validation data subsets within that node.
  - o  *Node Id:* This simple identifier makes it easy to identify the exact node.
  - o  *Percentage:* The percentage of the target variable represents DMR customers who have more than one subscription.
  - o  *Count:* The count is the number of DMR customers in that node.

## Node Interpretation

In Output 8.5, the topmost node on the left represents the whole sample of 8,010 DMR customers. Both the Train and Validation data subsets show that approximately 73% of DMR customers have more than a single subscription (MULTI_SUBS > 1) as shown in Output 8.5.

The first-level split is on HOUSEHOLD INCOME at $28,500. The upper box (Node 2) is lighter and represents 2,185 DMR customers with income less than $28,500. Within Node 2, 47.7% of DMR customers have more than one subscription. The lower box (Node 3) is darker and represents customers with annual income greater than $28,500. The percentage of DMR customers with more than a single subscription (MULTI_SUBS > 1) is 82.5%.

The second-level splits on a different variable for each node in the first level. Node 2 splits on AGE into Node 4 and Node 5. Node 4 shows that 81.4% of DMR customers younger than 22.5 years have more than one subscription. Node 5 shows that only 41.8% of DMR customers who are older than 22.5 years have more than one subscription.

The second split on the second level (Node 3) is on Household Income. Node 6 shows that 75.1% of DMR Customers with less than $51,500 in annual income have more than one subscription. Node 7 shows that 87.2% of DMR Customers with more than $51,500 in income have more than one subscription.

In Output 8.4, the final nodes show a wide range of percentages for DMR Customers having more than one subscription. Node 10 is the lightest background and has the lowest percentage of DMR Customers with more than one subscription (30.0%). Nodes 12 and 14 are the darkest, with the highest percentages of DMR Customers with more than one subscription (90.8% and 97.9%, respectively).

**NOTE:** "Or Missing" shows up in the bottom brackets whether or not you have missing values. In the current data, you have no missing values, so you can ignore "Or Missing" in this interpretation.

**Output 8.5: View Horizontal Tree End Points**

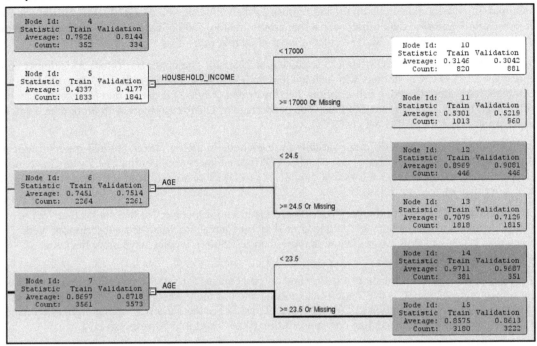

---

## Interpret the Findings

The results of this analysis offer meaningful insights into the demographic characteristics of DMR customers with more than one subscription. If you consider DMR Customers with multiple subscriptions to be your best customers, there are several approaches you can take:

- *Buy more prospect names:* You may decide to buy names from a list company with the same demographics as in the highest performing nodes. So for example, Node 12 and Node 14 identify the best performing customers. Nodes 13 and 15 identify reasonably high performing customers. So, you would look to purchase names that have the characteristics of Nodes 13 and 15.

- *Target your worst customers:* Based on the known demographics of the worst performing customers, you might want to create additional products that would meet their needs. Node 10 represents a group of customers that show untapped potential. Developing and testing new products might prove to be a good business decision.

- *Leverage your best customers:* These customers are already doing a lot of business with you. A simple survey of their other interests or preferred delivery channels may open new opportunities to increase profits within this customer group.

Your next question may be, "How can I precisely identify the customers in these nodes?" Understanding how to define your subscribers using the **Node Rules** is the first step.

The **Node Rules** define each node in simple language based on the characteristics of the customers within a specific node. Within the Results window, go to **View ▶ Model ▶ Node Rules** (Figure 8.16).

**Figure 8.16: Access and View Node Rules**

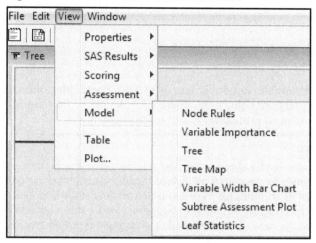

In Output 8.6, the **Node Rules** for Node 14 are shown. Node 14 has the highest percentage of DMR multi-subscribers.

**Output 8.6: View Node Rules—Highest Performing Node**

```
51    *------------------------------------------------------------------*
52     Node = 14
53    *------------------------------------------------------------------*
54    if HOUSEHOLD_INCOME >= 51500 or MISSING
55    AND AGE < 23.5
56    then
57     Tree Node Identifier   = 14
58     Number of Observations = 381
59     Predicted: MULTI_SUBS = 0.9711286089
60
```

Node 14 customers, with a 97.1% rate of multiple subscriptions, can be described as having an annual HOUSEHOLD_INCOME greater than $51,500 and being younger than 23.5 years. This group of customers seems to represent young men and women in affluent households. Using these very specific demographic characteristics, you can purchase more prospect names that have similar characteristics.

Another view of the nodes with the lowest multiple-subscriber rate can give you an idea of how to grow your business. In Output 8.7, you can learn the characteristics of these low-performing groups.

**Output 8.7: View Node Rules—Lowest Performing Node**

```
11    *----------------------------------------------------------*
12      Node = 10
13    *----------------------------------------------------------*
14    if HOUSEHOLD_INCOME < 17000
15    AND AGE >= 22.5 or MISSING
16    then
17      Tree Node Identifier   = 10
18      Number of Observations = 820
19      Predicted: MULTI_SUBS = 0.3146341463
20
```

Node 10 is the lowest performing group, with a multi-subscriber rate of 31.5%. These subscribers are characterized as low income (<$17,000 per year) and being older than 22.5 years. Some new marketing or products tailored for this group could prove to be profitable.

## Alternate Uses for Tree Analysis

Tree analysis can be used for a variety of goals. In this chapter, the tree was used as a descriptive analysis. In Chapter 9, you will use the tree node to develop a predictive model with a binary outcome. A third use of tree analysis is to create variables as an input for a regression or neural network model. Trees are also useful for diagnosing and interpreting the meaning of neural networks and other advanced modeling techniques. Simply point the tree at the neural network predicted target and grow a tree. For more information about this third use, see *Decision Trees for Analytics Using SAS Enterprise Miner*, by Barry de Ville and Padraic Neville (2013).

## Notes from the Field

With the influx of big data, the demand for predictive analytics continues to grow across a variety of industries and applications. From marketing and risk to process improvement and human resources, the possibilities for applying tree models are endless. Your ability to build powerful tree models will enable you to provide meaningful insights and direction to stakeholders across your enterprise.

As the person with the keys to the Ferrari, you play an important role in shaping the analysis. In other words, you have access to advanced techniques within SAS software that enable you to analyze big data and provide insights and direction to your clients and stakeholders. But, because most of your clients and stakeholders don't understand what you do or how you do it, you must be very diligent about your process from beginning to end. At each step, ask yourself if what you are seeing makes sense in terms of your business. And, remember to drive carefully.

# Chapter 9: Predictive Analysis Using SAS Enterprise Miner

## Introduction

In Chapters 5 through 8, you learned to view and analyze the DMR Publishing data in an effort to understand past and current customer characteristics and behavior. In this chapter, you will learn how to discover and measure patterns from past behavior and characteristics to predict future behavior using predictive modeling. This is one of most widely used techniques used in business today.

As detailed in Chapter 1, there are numerous uses for predictive modeling in marketing, risk, process improvement, customer retention, and more. In this chapter, you will build a model to predict the likelihood of loan default using credit risk data.

To achieve your goal, you will initiate a project in SAS Enterprise Miner and proceed through the model building process. You will explore and manipulate the data to prepare for modeling. You will build three common types of models: a decision tree, a neural network, and a regression model. You will compare the three models in an assessment node.

The steps for model development are as follows:

- *Select:* Initiate the project and bring the modeling data set into the project. Be sure that it correlates with the business purpose for building the model.
- *Explore:* Use visual tools to view the distribution and completeness of the data.
- *Modify:* Partition the data into training, test, and validation data sets. Impute missing values and filter outliers. Transform or segment interval variables if necessary. Create indicator variables for variables with character or non-sequential values.
- *Model:* Develop three different models: a decision tree, a neural network, and a regression model.
- *Assess*: Compare the three models by using the gains table and lift charts.

> **NOTE:** For supporting documentation on every aspect of SAS Enterprise Miner, see the Contents section under Help in the main menu.

## Select

In this section, you initiate a project in SAS Enterprise Miner and select data for the model-building process.

### Initiate the Project

ABC Bank provided a "snapshot" of customer data that includes customers who defaulted on their loans and customers who did not default on their loans. The customer payment, balance, and several demographic variables were appended from a credit bureau for a period six months earlier. The dependent variable is a field called LOAN_DEFAULT. The goal of the model is to learn whether you can predict who is most likely to default on a loan six months from now.

Open SAS Enterprise Miner and select **New Project**. Name your project and select a place to store the files as seen in Figure 9.1. Click **Next ▶ Finish**.

**Figure 9.1: Initiate a New Project**

## Select the Data

Open the Data Source Wizard by clicking the **Create Data Source** icon shown in Figure 9.2. We are searching for the ABC Bank data in the Sasuser folder.

**Figure 9.2: Use the Data Source Wizard**

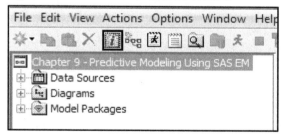

When the window opens, click **Next ▶ Browse** and select Sasuser as shown in Figure 9.3.

**Figure 9.3: Select Sasuser Folder**

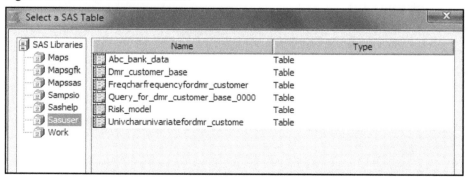

Double-click the Sasuser folder and highlight ABC_Bank_Data. Click **OK** ▶ **Next** ▶ **Next** ▶ **Next** and you will be at **Step 5 of 8**. Click the button next to **Advanced** and then click **Next**. In this step, you assign the dependent variable.

The goal of the project is to build a model to predict loan default, so, the LOAN_DEFAULT variable is the dependent variable for the project. Next to the Name LOAN_DEFAULT, change the **Role** to **Target** and the **Level** to **Binary** (if it is not Binary by default) as shown in Figure 9.4.

**Figure 9.4: Assign Dependent Variable**

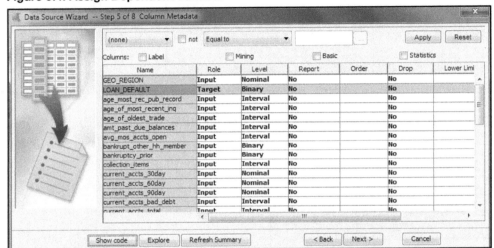

Click **Next** ▶ **Next** ▶ **Next** ▶ **Finish**. The data set is now ready to begin the modeling process.

Next, create a new diagram. Click **File** ▶ **New** ▶ **Diagram** and enter a name as shown in Figure 9.5. Click **OK**.

**Figure 9.5: Name a New Diagram**

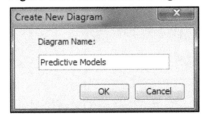

Once the diagram appears, drag the ABC_Bank_Data into the diagram as shown in Figure 9.6.

**Figure 9.6: View the Project and the Data Source in the Modeling Diagram**

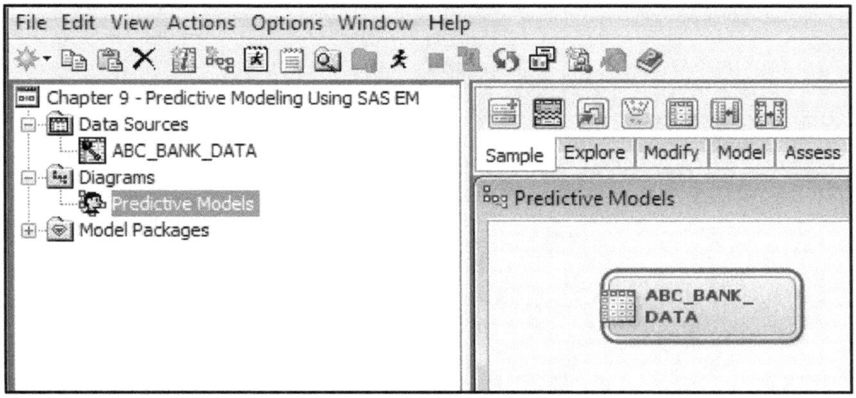

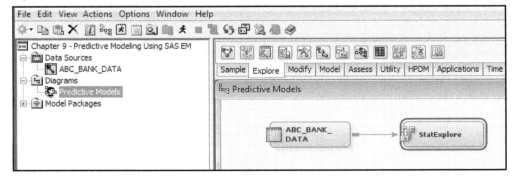

## Explore

When developing a predictive model, you need to prepare the data for modeling, so you may need to take steps to handle missing values and outliers. You also want the relationship between the dependent variable and the independent variables to follow certain distributions.

To view the data, use the StatExplore and MultiPlot nodes. These nodes allow you to view the distributions of the variables, as well as their relationships to the target variable.

### StatExplore

To initiate StatExplore, in the **Explore** tab look for the icon (third from the right), drag it onto the diagram, and connect it to ABC_Bank_Data with an arrow as seen in Figure 9.7.

**Figure 9.7: Project View with the Data Source in the Modeling Diagram**

Right-click the StatExplore icon; click **Run ▶ Yes**. When the run completes, select **Results**.

**Output 9.1: StatExplore Results**

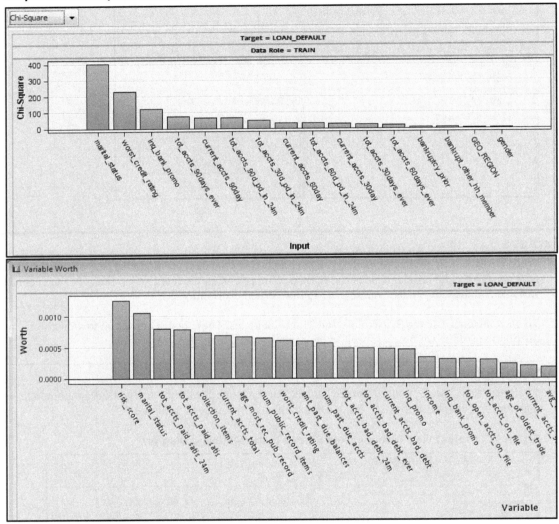

The upper plot is the chi-square plot. It shows the strength of the relationship between the variables and the dependent variable. Here, marital_status has the strongest correlation with the target variable, LOAN_DEFAULT.

The Variable Worth plot measures and displays the independent variables according to their calculated worth. RISK_SCORE has the highest worth, with marital_status coming in second, tot_accts_paid_satis_last_24m coming in third, and so on. For more information on variable worth, see the variable importance explanation in the decision tree section of the Contents within SAS Enterprise Miner. You can access Contents by clicking **Help ▶ Contents** on the upper menu.

The upper right window of the Results window offers a rich set of statistics on each variable. Enlarge the window and scroll down to see various features of the variables. In Output 9.2, you get a partial view that shows the class variables. Notice that marital_status has 924 missing values. These missing values will be imputed in the upcoming "Modify" section, using the impute process.

**Output 9.2: StatExplore Statistics—Class Variables**

| | | | | | | | | | | |
|---|---|---|---|---|---|---|---|---|---|---|
| 34 | Class Variable Summary Statistics | | | | | | | | | |
| 35 | (maximum 500 observations printed) | | | | | | | | | |
| 36 | | | | | | | | | | |
| 37 | Data Role=TRAIN | | | | | | | | | |
| 38 | | | | | | | | | | |
| 39 | | | | | Number | | | | Mode | | Mode2 |
| 40 | Data | | | | of | | | Mode | | Mode2 | |
| 41 | Role | Variable Name | | Role | Levels | Missing | Mode | Percentage | Mode2 | Percentage |
| 42 | | | | | | | | | | |
| 43 | TRAIN | GEO_REGION | | INPUT | 4 | 0 | West | 36.95 | Midwest | 27.59 |
| 44 | TRAIN | bankrupt_other_hh_member | | INPUT | 2 | 0 | 0 | 99.74 | 1 | 0.26 |
| 45 | TRAIN | bankruptcy_prior | | INPUT | 2 | 0 | 0 | 91.89 | 1 | 8.11 |
| 46 | TRAIN | current_accts_30day | | INPUT | 8 | 0 | 0 | 88.91 | 1 | 8.77 |
| 47 | TRAIN | current_accts_60day | | INPUT | 7 | 0 | 0 | 94.31 | 1 | 4.79 |
| 48 | TRAIN | current_accts_90day | | INPUT | 14 | 0 | 0 | 89.01 | 1 | 7.80 |
| 49 | TRAIN | gender | | INPUT | 3 | 0 | M | 52.23 | F | 43.98 |
| 50 | TRAIN | inq_bank_promo | | INPUT | 17 | 0 | 0 | 42.73 | 1 | 21.58 |
| 51 | TRAIN | marital_status | | INPUT | 5 | 924 | M | 32.53 | S | 30.03 |
| 52 | TRAIN | tot_accts_30d_pd_in_24m | | INPUT | 13 | 0 | 0 | 76.55 | 1 | 15.07 |
| 53 | TRAIN | tot_accts_30days_ever | | INPUT | 15 | 0 | 0 | 66.21 | 1 | 18.36 |
| 54 | TRAIN | tot_accts_60d_pd_in_24m | | INPUT | 10 | 0 | 0 | 88.95 | 1 | 8.37 |
| 55 | TRAIN | tot_accts_60days_ever | | INPUT | 10 | 0 | 0 | 82.38 | 1 | 12.19 |
| 56 | TRAIN | tot_accts_90d_pd_in_24m | | INPUT | 16 | 0 | 0 | 85.33 | 1 | 9.48 |
| 57 | TRAIN | tot_accts_90days_ever | | INPUT | 16 | 0 | 0 | 78.58 | 1 | 12.52 |
| 58 | TRAIN | worst_credit_rating | | INPUT | 7 | 0 | 1 | 64.41 | 6 | 22.83 |
| 59 | TRAIN | LOAN_DEFAULT | | TARGET | 2 | 0 | 0 | 95.84 | 1 | 4.16 |
| 60 | | | | | | | | | | |

Output 9.3 shows that the independent variable "income" has 549 missing values. We will populate these in the upcoming "Modify" section using the impute process.

## Output 9.3: StatExplore Statistics—Interval Variables

```
76   Interval Variable Summary Statistics
77   (maximum 500 observations printed)
78
79   Data Role=TRAIN
80
```

| | | | | Standard | Non | | | | | | |
| | Variable | Role | Mean | Deviation | Missing | Missing | Minimum | Median | Maximum | Skewness | Kurtosis |
|---|---|---|---|---|---|---|---|---|---|---|---|
| 84 | age_most_rec_pub_record | INPUT | 5.762059 | 14.54792 | 28734 | 0 | 0 | 0 | 119 | 3.572219 | 14.41568 |
| 85 | age_of_most_recent_inq | INPUT | 4.551681 | 5.775763 | 28734 | 0 | 0 | 2 | 24 | 1.527642 | 1.530631 |
| 86 | age_of_oldest_trade | INPUT | 108.3721 | 90.67497 | 28734 | 0 | 0 | 86 | 829 | 1.486329 | 2.536665 |
| 87 | amt_past_due_balances | INPUT | 698.3108 | 7700.961 | 28734 | 0 | 0 | 0 | 1119197 | 113.1366 | 15791.41 |
| 88 | avg_mos_accts_open | INPUT | 52.02311 | 36.16629 | 28734 | 0 | 0 | 46 | 469 | 1.756677 | 6.893803 |
| 89 | collection_items | INPUT | 0.849655 | 2.411677 | 28734 | 0 | 0 | 0 | 50 | 6.109127 | 59.09067 |
| 90 | current_accts_bad_debt | INPUT | 0.844574 | 2.134596 | 28734 | 0 | 0 | 0 | 45 | 4.317058 | 29.69401 |
| 91 | current_accts_total | INPUT | 8.464711 | 7.932097 | 28734 | 0 | 0 | 6 | 75 | 1.532718 | 3.248087 |
| 92 | income | INPUT | 21034.03 | 23446.74 | 28185 | 549 | 0 | 14246 | 325183 | 2.414008 | 10.55158 |
| 93 | inq_fin_last_6mos | INPUT | 0.753706 | 1.722227 | 28734 | 0 | 0 | 0 | 38 | 5.214959 | 49.05728 |
| 94 | inq_past_12mos | INPUT | 1.124104 | 1.669604 | 28734 | 0 | 0 | 1 | 26 | 2.894731 | 15.42647 |
| 95 | inq_promo | INPUT | 3.361941 | 3.343572 | 28734 | 0 | 0 | 2 | 27 | 1.4799 | 2.96967 |
| 96 | num_past_due_accts | INPUT | 0.582098 | 1.392362 | 28734 | 0 | 0 | 0 | 41 | 5.148837 | 61.8276 |
| 97 | num_public_record_items | INPUT | 1.182432 | 3.00794 | 28734 | 0 | 0 | 0 | 65 | 5.463535 | 47.98795 |
| 98 | risk_score | INPUT | 808.7328 | 155.6376 | 28734 | 0 | 222 | 863 | 994 | -1.07662 | 0.314598 |
| 99 | tot_accts_bad_debt_24m | INPUT | 0.611506 | 1.713559 | 28734 | 0 | 0 | 0 | 45 | 5.299255 | 49.85442 |
| .00 | tot_accts_bad_debt_ever | INPUT | 0.854528 | 2.154509 | 28734 | 0 | 0 | 0 | 45 | 4.293498 | 29.21094 |
| .01 | tot_accts_on_file | INPUT | 11.10138 | 8.838177 | 28734 | 0 | 1 | 9 | 81 | 1.390275 | 2.815527 |
| .02 | tot_accts_open_last_24m | INPUT | 2.772186 | 2.632364 | 28734 | 0 | 0 | 2 | 42 | 1.73403 | 6.044214 |
| .03 | tot_accts_paid_satis | INPUT | 7.539361 | 7.615784 | 28734 | 0 | 0 | 5 | 75 | 1.668434 | 3.840586 |
| .04 | tot_accts_paid_satis_24m | INPUT | 5.843844 | 5.720464 | 28734 | 0 | 0 | 4 | 64 | 1.65111 | 4.258955 |
| .05 | tot_bal_open_accts | INPUT | 10885.18 | 14350.11 | 28734 | 0 | 0 | 5988 | 251742 | 3.159175 | 20.69761 |
| .06 | tot_open_accts_bal_gt_0 | INPUT | 3.580358 | 2.994385 | 28734 | 0 | 0 | 3 | 46 | 1.880609 | 7.488173 |
| .07 | tot_open_accts_on_file | INPUT | 5.394028 | 4.238754 | 28734 | 0 | 0 | 4 | 51 | 1.543187 | 4.295503 |
| .08 | | | | | | | | | | | |

Close the **Results** window by clicking the X in the upper right corner.

## MultiPlot

The MultiPlot node allows you to examine the relationships between the variables, as well as their distributions. This ability is useful for determining the univariate power of an independent variable to predict the dependent variable. It can also help you determine the best transformation, if necessary.

To initiate MultiPlot, find the icon in the Explore tab (seventh from the left), drag it onto the diagram, and connect it to the ABC_Bank_Data with an arrow.

Check the Properties window on the left-hand side to ensure that the **Type of Charts** is set to **Bar Charts** as shown in Figure 9.8.

**Figure 9.8: MultiPlot Properties Window**

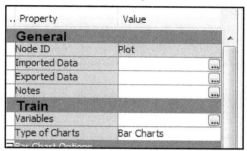

Right-click the icon and click **Run ▶ Yes**. When the run completes, click **Results**. The window opens with two sections. Expand the top panel to see the bar charts.

This process creates a frequency bar chart for each independent variable by LOAN_DEFAULT. The bar chart for tot_bal_open_Accts by LOAN_DEFAULT is shown in Output 9.4. The menu at the bottom of the chart enables you to look at the relationship between all the independent variables and LOAN_DEFAULT. Click the center arrow to start a slide show of the different frequencies. Or, use the right and left arrows to view at your own pace.

**Output 9.4: MultiPlot Total Balance on Open Accounts by Loan Default—Bar Chart**

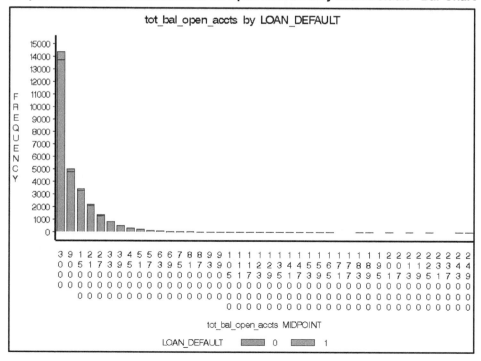

These graphs offer a quick view of the distribution of each independent variable in relation to LOAN_DEFAULT. The variable is also highly skewed to the right. In subsequent steps, you will scrub the tot_bal_open_Accts variable for outliers and transform it to make it more normally distributed.

# Modify

This section illustrates how to prepare the data for modeling. You will impute values to replace missing values, partition the data, filter outliers, and transform the independent variables.

## Replace Missing Values via Imputation

The next step is to impute values for missing or incorrect observations. Recall that the StatExplore window revealed that the Marital Status and Income variables have missing values.

Begin by clicking **Modify** and dragging the **Impute** icon onto the diagram, as shown in Figure 9.9.

**Figure 9.9: Impute Missing Values**

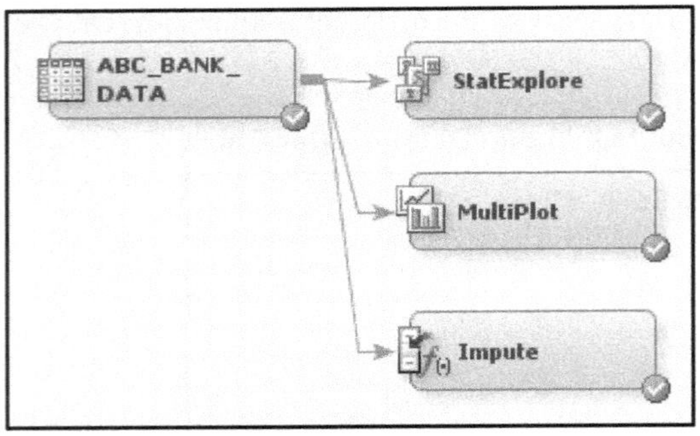

The Property window is populated with default methods for imputing missing values. The class variables are replaced with the value that has the highest count. The highest count value is used as a replacement for Marital Status. Recall that Gender has a value = U for *unknown*. You can leave this value as is: Sometimes the fact that the Gender is unknown is predictive. If it were missing, it would be set to M because of a higher count of customers with Gender = M.

For interval variables, the default method is to replace missing values with the mean value. Because your number of missing values is small, this default is adequate. If it were a larger percentage, then using a more advanced method, such as tree, might be better. To learn more about the Tree Imputation Method, see Contents in the Help section of Enterprise Miner.

Right-click on the **Impute** icon and click **Run ▶ Yes**. When complete, click **Results**. Output 9.5 shows a partial view of the number of imputed values as well as the value used to replace the missing values.

**Output 9.5: Impute Missing Value Results**

| Variable Name | Impute Method | Imputed Variable | Impute Value |
|---|---|---|---|
| ncome | MEAN | IMP_income | 21034.032358 |
| marital_status | COUNT | IMP_marital_status | M |

## Partition Data into Subsamples

The next step is to partition the data into three subsamples: Train, Validate, and Test. In your diagram, click the **Sample** tab, drag the **Partition** icon onto the diagram, and connect it to the **Impute** icon as shown in Figure 9.10.

**Figure 9.10: Partition Data**

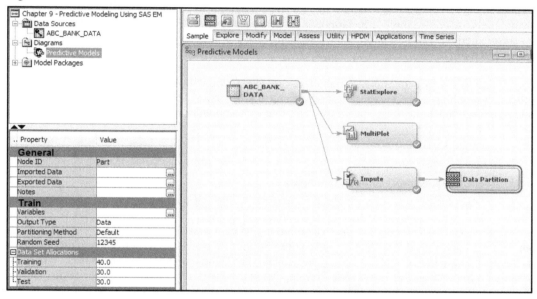

For every modeling technique, the **Train** data is the base data set used to build several interim models and a final model during the model building process. The **Validation** data is used to fine tune each model and avoid overfitting during the decision tree, neural network, and stepwise regression model building processes. The **Test** data is used to evaluate the final model on unbiased data not used in the model building process. The final model is the one with the best fit based on your preselected criteria.

You will build three models, one each of decision tree, neural network, and stepwise regression. These will be evaluated individually and also compared using the **Model Comparison** node.

Notice in the Property window at the left that the partitioning method, under **Train**, is set to **Default**. The choices are **Simple, Random, Cluster**, and **Stratified**. You want to use the **Stratified** method, and because you have a binary target variable, the default is **Stratified.** So, you don't need to make any change here. If you do change it to **Stratified**, the result will be the same.

Notice the default percentages in the **Property** window. Under **Data Set Allocations**, the default settings assign 40% to training, 30% to validation, and 30% to test. These defaults can be changed as long as the total equals 100%. Right-click on the Data Partition process and click **Run ▶ Yes**. You can view the results to see that the percentage of the dependent variable is close in each of the three sample data sets.

## Manage Outliers

As described in Chapter 5, outliers are data points that are unusually large or small when compared to the rest of the values for a certain variable. While there is no precise rule to identify an outlier, you can look for values that are more than 3, 4, or 5 standard deviations from the mean.

Outliers may also represent data errors. Or, for some modeling applications, outliers may be very useful to your analysis. For example, when modeling a rare event such as fraud, you might find the outliers to be very predictive. The best approach to modeling rare events is using neural networks, which are less sensitive to outliers.

When building a regression model for a not-so-rare event, you will want to take steps to manage the outliers. Depending on the number of observations in your data set compared to the number of outliers, you can choose to handle them differently. The four most common methods for dealing with outliers are:

1. Cap the variable.
2. Drop the observations.
3. Treat them as errors or missing values.
4. Bucket the continuous variable into quartiles, deciles, or centiles.

Regression models are especially sensitive to outliers. If your data contains outliers, investigating the source of the data is good practice. Because your current data comes directly from a credit bureau and has not been altered to correct for outliers, it typically has some outliers. In this case study, you will filter the outliers. In practice, this is a simple first step using the Filter node. If you have the luxury of testing different methods for handling outliers, you can compare the model results and develop your own preference for managing outliers.

To filter outliers, click on the **Sample** tab and drag the **Filter** icon to your diagram, as shown in Figure 9.11. In the **Property** window, set the **Tables to Filter** to **All Data Sets.** The default setting for **Class Variables** is **Rare Values (Percentage).** Leave this setting as is. For **Interval Variables**, set the method to **Extreme Percentiles**. The **Extreme Percentiles** method filters out the top 5% of values. This setting can be changed by clicking the three dots to the right of **Tuning Parameters**.

**Figure 9.11: Filter Extreme Values**

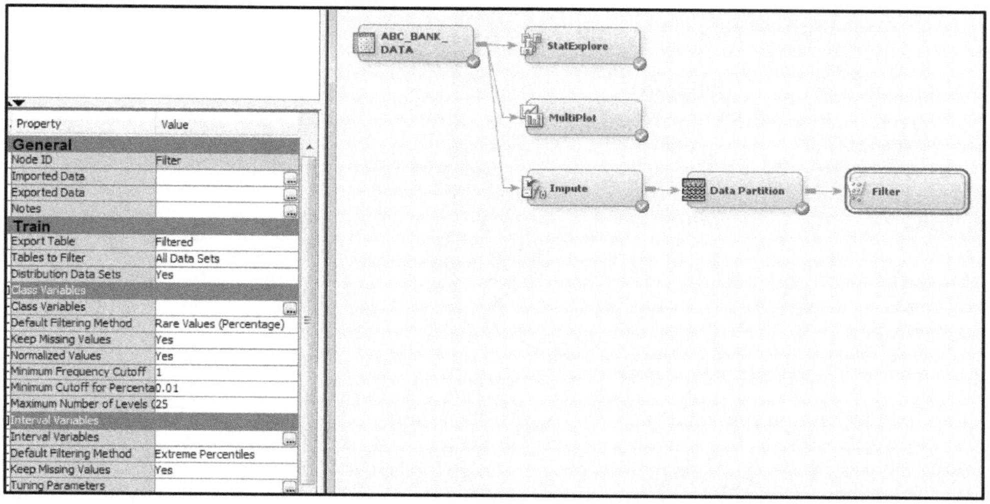

Right-click the **Filter** icon, then click **Run ▶ Yes.** The overview of the filter results is shown in Output 9.6.

**Output 9.6: Overview of the Filter Results**

```
144
145    Number Of Observations
146
147    Data
148    Role        Filtered    Excluded    DATA
149
150    TRAIN        10044        1447      11491
151    VALIDATE      7487        1133       8620
152    TEST          7493        1130       8623
153
```

Output 9.7 displays the changes in values for amt_past_due_balances for the TEST data set. Notice that the total number of records for amt_past_due_balances dropped from 8,623 to 7,487. The maximum dropped from 1,119,197 to 18,022.

**Output 9.7: Filter Results—Amount of Past Due Balances**

```
215    Data Role=TEST Variable=amt_past_due_balances
216
217    Statistics                Original     Filtered
218
219    Non Missing                8623.00      7487.00
220    Missing                       0.00         0.00
221    Minimum                       0.00         0.00
222    Maximum                 1119197.00     18022.00
223    Mean                        739.85       431.96
224    Standard Deviation        12351.61      1597.30
225    Skewness                     86.27         5.53
226    Kurtosis                   7801.03        36.35
227
```

The next step is to transform the variables.

> **NOTE:** If you have hundreds of variables, you might want to use the variable selection node. For details about how to use the variable selection node, see the Help section of SAS Enterprise Miner.

## Transform the Variables

In this section, you transform the distribution of the variables to create a better fit for the neural network and regression models. From the **Modify** tab, drag the **Transform Variables** icon (far right) onto the diagram and connect it to the Filter icon. Highlight the **Transform Variables** icon to unveil the Property window.

As shown in the **Property** window in Figure 9.12, change the default methods as follows: For **Interval Inputs,** select **Best**. This method tries several transformations and selects the best fit. For the **Class Inputs,** select **Dummy Indicators**. This option gives each level of the class variables values of 0 and 1.

> **NOTE**: The diagram in Figure 9.12 has been rearranged to accommodate space. If you want to rearrange your diagram, just click and drag the icons to the arrangement that works for you.

**Figure 9.12: Transform Variables Property Window**

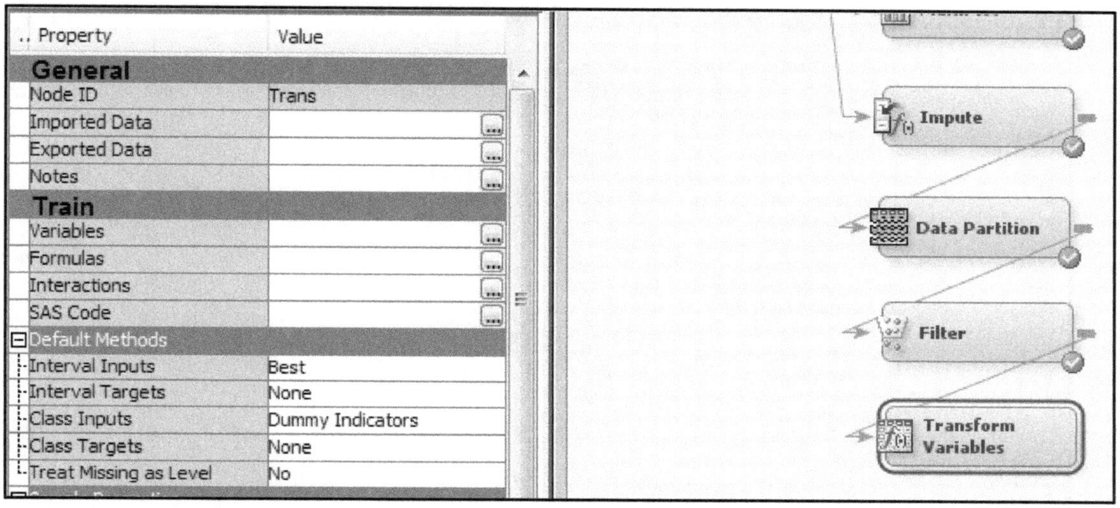
.. Property
Value
General
Node ID
Trans
Imported Data
Exported Data
Notes
Train
Variables
Formulas
Interactions
SAS Code
Default Methods
Interval Inputs
Best
Interval Targets
None
Class Inputs
Dummy Indicators
Class Targets
None
Treat Missing as Level
No
Impute
Data Partition
Filter
Transform
Variables

Right-click the Transform Variables icon, and then click **Run ▶ Yes**. When the transform process is complete, click **Results**.

**Output 9.8: Transform Variables Results**

| Source | Method | Variable Name | Formula | Number of Levels | Non Missing | Missing | Minimum | Maximum | Mean |
|--------|--------|---------------|---------|------------------|-------------|---------|---------|---------|------|
| Input | Original | GEO_REGI... | | 4 | | 0 | . | . | . |
| Input | Original | IMP_income | | | 10044 | 0 | 0 | 123548 | 18751.84 |
| Input | Original | IMP_marital... | | 4 | | 0 | . | . | . |
| Input | Original | age_most_... | | | 10044 | 0 | 0 | 83 | 5.139586 |
| Input | Original | age_of_mo... | | | 10044 | 0 | 0 | 23 | 4.420251 |
| Input | Original | age_of_old... | | | 10044 | 0 | 2 | 439 | 99.69215 |
| Input | Original | amt_past_... | | | 10044 | 0 | 0 | 18225 | 403.5974 |
| Input | Original | avg_mos_a... | | | 10044 | 0 | 2 | 212 | 48.8636 |
| Input | Original | bankrupt_ot... | | 1 | | 0 | . | . | . |
| Input | Original | bankruptcy... | | 2 | | 0 | . | . | . |
| Input | Original | collection_it... | | | 10044 | 0 | 0 | 16 | 0.700119 |
| Input | Original | current_acc... | | 3 | | 0 | . | . | . |
| Input | Original | current_acc... | | 2 | | 0 | . | . | . |
| Input | Original | current_acc... | | 3 | | 0 | . | . | . |
| Input | Original | current_acc... | | | 10044 | 0 | 0 | 13 | 0.661987 |
| Input | Original | current_acc... | | | 10044 | 0 | 0 | 38 | 7.66816 |
| Input | Original | gender | | 3 | | 0 | . | . | . |
| Input | Original | inq_bank_p... | | 7 | | 0 | . | . | . |
| Input | Original | inq_past_1... | | | 10044 | 0 | 0 | 9 | 1.042413 |

Best Transformation R Square

| Original Variable ▼ | Computed Variable | Formula |
|---------------------|-------------------|---------|
| tot_open_accts_on_file | INV_tot_open_accts_on_file | 1 / (tot_open_accts_on_file + 1) |
| tot_open_accts_on_file | OPT_tot_open_accts_on_file | Optimal Binning(4) |
| tot_open_accts_on_file | LOG_tot_open_accts_on_file | log(tot_open_accts_on_file + 1) |
| tot_open_accts_on_file | LG10_tot_open_accts_on_file | log10(tot_open_accts_on_file + 1) |
| tot_open_accts_on_file | SQRT_tot_open_accts_on_file | Sqrt(tot_open_accts_on_file + 1) |
| tot_open_accts_on_file | STD_tot_open_accts_on_file | (tot_open_accts_on_file - 4.9193548387) / 3.6472547393 |
| tot_open_accts_on_file | tot_open_accts_on_file | |
| tot_open_accts_on_file | CNTR_tot_open_accts_on_file | (tot_open_accts_on_file - 4.9193548387) |
| tot_open_accts_on_file | RANGE_tot_open_accts_on_file | (tot_open_accts_on_file - 0) / (22-0) |
| tot_open_accts_on_file | SQR_tot_open_accts_on_file | (tot_open_accts_on_file + 1)**2 |
| tot_open_accts_on_file | EXP_tot_open_accts_on_file | exp(tot_open_accts_on_file ) |
| tot_open_accts_bal_gt_0 | INV_tot_open_accts_bal_gt_0 | 1 / (tot_open_accts_bal_gt_0 + 1) |
| tot_open_accts_bal_gt_0 | OPT_tot_open_accts_bal_gt_0 | Optimal Binning(4) |
| tot_open_accts_bal_gt_0 | LOG_tot_open_accts_bal_gt_0 | log(tot_open_accts_bal_gt_0 + 1) |
| tot_open_accts_bal_gt_0 | LG10_tot_open_accts_bal_gt_0 | log10(tot_open_accts_bal_gt_0 + 1) |
| tot_open_accts_bal_gt_0 | SQRT_tot_open_accts_bal_gt_0 | Sqrt(tot_open_accts_bal_gt_0 + 1) |
| tot_open_accts_bal_gt_0 | CNTR_tot_open_accts_bal_gt_0 | (tot_open_accts_bal_gt_0 - 3.2114695341) |
| tot_open_accts_bal_gt_0 | RANGE_tot_open_accts_bal_gt_0 | (tot_open_accts_bal_gt_0 - 0) / (16-0) |

The lower window shows all the transformations that were created for each interval variable. Expand to see the full list. These transformations will be considered as inputs to the neural network and regression modeling processes.

Now you are ready to build, validate, and test your models.

# Model

Now the fun really begins. The first model is a decision tree. Because the decision tree process finds discrete groupings to build the model, this modeling process will not use the transformed variables.

## Decision Tree

In Chapter 8, you built a decision tree to describe the data. In this chapter, you are building a decision tree to predict the likelihood of an event occurring. This method is versatile and can be used in multiple ways.

> **NOTE**: Recall that you replaced the missing values in your data using the Impute function. If you had not replaced these values, the Decision Tree process would treat them as a separate category. See Enterprise Guide Help Contents for more information.

Under the **Model** tab, drag the **Decision Tree** icon onto the diagram and connect it to the **Filter** node as shown in Figure 9.13. Two changes are recommended in the **Property** window. Under **Splitting Rule**, change the **Maximum Branch** to 4. This option is something you can explore, going higher and lower. Under **Subtree**, change the **Method** to **Assessment** and the **Assessment Measure** to **Lift**. Right-click the **Decision Tree** icon and click **Run ▶ Yes**.

**Figure 9.13: Decision Tree Property Window**

| ., Property | Value |
|---|---|
| Missing Values | Use in search |
| Use Input Once | No |
| Maximum Branch | 4 |
| Maximum Depth | 6 |
| Minimum Categorical Size | 5 |
| **Node** | |
| Leaf Size | 5 |
| Number of Rules | 5 |
| Number of Surrogate Rules | 0 |
| Split Size | . |
| **Split Search** | |
| Use Decisions | No |
| Use Priors | No |
| Exhaustive | 5000 |
| Node Sample | 20000 |
| **Subtree** | |
| Method | Assessment |
| Number of Leaves | 1 |
| Assessment Measure | Lift |
| Assessment Fraction | 0.25 |

Once the run completes, click **Results**. The graph in the upper left corner of the output window shows the cumulative lift graph as seen in Output 9.9. The graph shows that the lift in the train data is much better than the lift in the validation data.

**Output 9.9: Decision Tree Results**

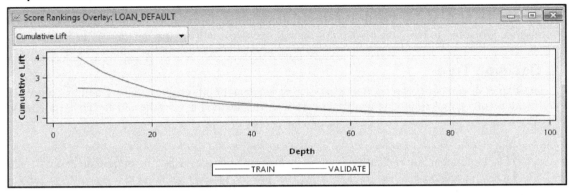

The **Train** data appears to capture 35% of the defaults in the top 10% of the file. The **Validation** data captures around 25% of the defaults in the top 10% of the file. The performance of the model on both the **Train** and **Validation** data shows an improvement. You will compare the results of the decision tree model with the results of the neural network and regression models in subsequent sections.

The partial decision tree view seen in Output 9.10 shows that RISK_SCORE is the strongest predictor. This result matches the results from the interval variables in the StatExplore results, shown in Output 9.1.

**Output 9.10: Decision Tree View**

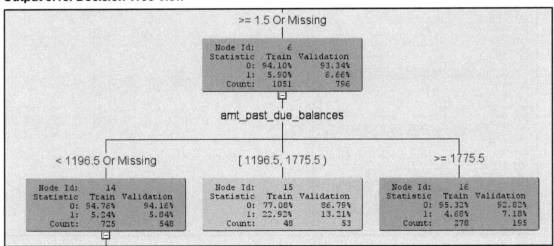

Because the decision tree is built for prediction, the actual tree node values are not as important as the model's ability to predict. Consequently, in the next few steps you will build the neural network and regression models.

In the final assessment step, you will employ the Model Comparison node to compare all three models, using both gains tables and a cumulative-lift graph.

## Neural Network

As described in Chapter 3, neural networks are models that capture many of the nonlinear relationships in data. So, although neural network models sometimes do a better job of fitting the model than the decision tree or regression models, they might not perform as well on the validation data or maintain their ability to predict as accurately over time. Another disadvantage of neural network models is that they are more difficult for marketers to interpret than the decision tree models.

Under the **Model** tab, drag the **Neural Network** icon (fifth from the right) onto the diagram and connect it to the **Transform Variables** node. Figure 9.14 shows the diagram and the **Neural Network Property** window. In the **Train** section, next to **Model Selection Criterion**, change **Value** to **Misclassification.** Right-click the model node, and click **Run ▶ Yes.** When the window appears, click **Results.**

**Figure 9.14: Neural Network Property Window**

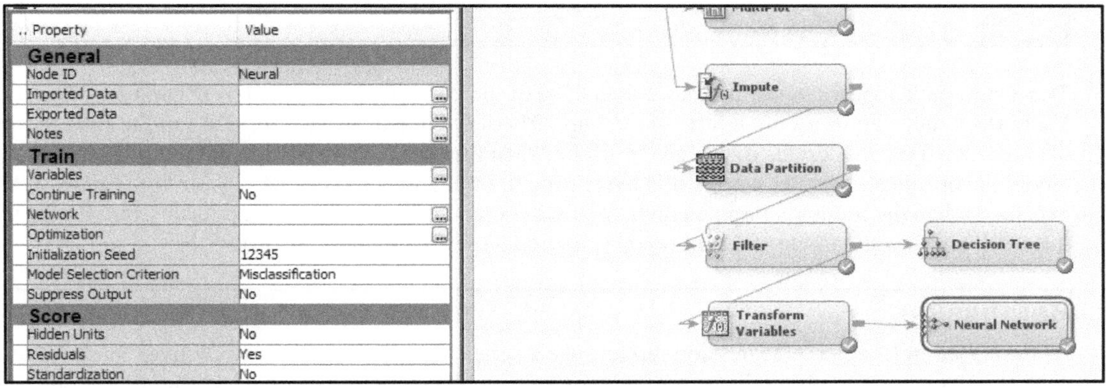

As shown in Output 9.11, the cumulative lift is much higher for the neural network train data than for the validation data.

**Output 9.11: Neural Network Cumulative Lift**

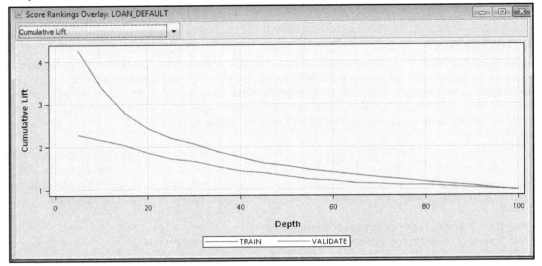

The **Train** data appears to capture more than 45% of the defaults in the top 10% of the file. The **Validation** data captures only a little more than 20% of the defaults in the top 10% of the file. As with the Decision Tree mode, this difference in lift between the train and validation data implies that the model is not robust. In other words, it is overfitting the train data and probably won't predict as well as expected when the model is implemented. You will compare the results of the decision tree model with the results of the neural network and regression models in an upcoming section.

## Regression

As described in Chapter 3, regression models use a series of independent, or predictive, variables to predict an outcome. In predictive modeling, you can use logistic regression, which builds a model with a binary outcome. The resulting equation assigns a probability of the event's occurring. In the current case, the binary outcome or event is **Loan Default**.

Under the **Model** tab, drag the **Regression** icon (third from the right) onto the diagram, and connect it to the **Transform Variables** node. Figure 9.15 shows the diagram and the **Regression Property** window. Because the target variable, LOAN_DEFAULT, is binary, the **Regression Type** defaults to **Logistic**.

As mentioned in Chapter 3, each selection option has pros and cons. For this model, change **Selection Model** to **Stepwise**. Change **Use Selection Defaults** to **No** (Figure 9.15).

**Figure 9.15: Regression Property Window**

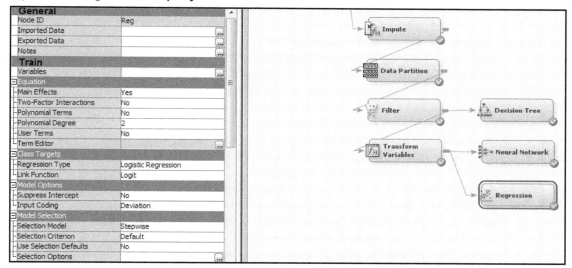

You might prefer the **Stepwise** selection mode with a slightly higher **Entry Significance Level** than the **Stay Significance Level**. This setting allows the model to have the benefit of the variables working together thereby avoiding one very powerful variable from dominating the model. It also works like the **Backward Selection Model** by removing variables that don't meet the **Stay Significance Level** after they are entered into the model. It can also continue to fine-tune the model by bringing in new variables that meet the significance threshold after other variables are eliminated. For more detail on selection options for regression, see *Data Mining Cookbook* (Parr-Rud 2001, 104-105).

Next, click the three dots to the right of **Selection Options.** The window (Figure 9.16) allows you to set the **Entry Significance Level** to 0.20 and the **Stay Significance Level** to 0.05. For **Maximum Number of Steps**, enter 20. This will allow for enough predictive variables to enter the model.

Right-click the **Regression** node and click **Run ▶ Yes.** When the window appears, click **Results.**

**Figure 9.16: Selection Options Significance Levels**

| .. Property | Value |
| --- | --- |
| Sequential Order | No |
| Entry Significance Level | 0.2 |
| Stay Significance Level | 0.05 |
| Start Variable Number | 0 |
| Stop Variable Number | 0 |
| Force Candidate Effects | 0 |
| Hierarchy Effects | Class |
| Moving Effect Rule | None |
| Maximum Number of Steps | 20 |

As shown in Output 9.12, the cumulative lift is much higher for the regression train data than for the validation data. The **Train** data appears to capture more than 30% of the defaults in the top 10% of the file. The **Validation** data captures only about 25% of the defaults in the top 10% of the file, which is the best performance of the **Validation** data so far.

**Output 9.12: Logistic Regression Cumulative Lift**

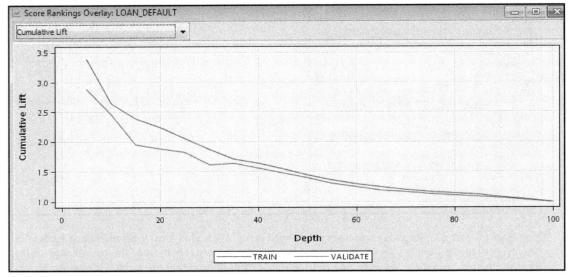

Next, you will compare the three models in the Model Comparison Node.

## Assess

The **Model Comparison** node allows you to visually compare the models and perform decile analysis using gains tables. From the menu at the top of the diagram, click the Assess tab. Select and drag the **Model Comparison** icon onto the diagram and connect it to all three models, as shown in Figure 9.17. In the **Property** window, change the **Number of Bins** to 10 to view the results in deciles and change **Selection Statistic** to **Cumulative Lift**. Right-click on the **Model Comparison** node and click **Run ▶ Yes**. When the window appears, click **Results**.

**Figure 9.17: Model Comparison Property Window**

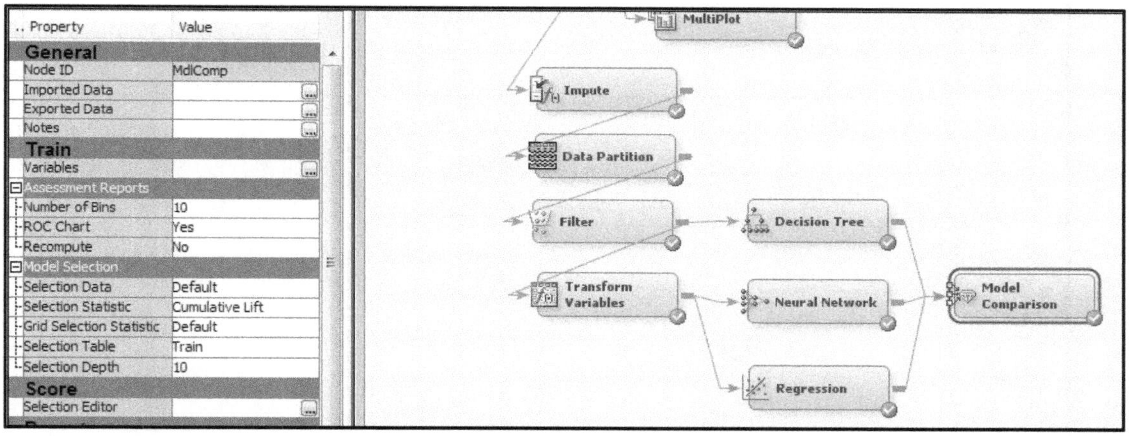

The lower left corner of the results window displays the cumulative lift graphs as seen in Output 9.13. All three models are compared for each of the three data sets: Train, Validate, and Test. The model performance for all three data sets shows the most lift in the top 10%. As expected, the neural network model and the decision tree have the best performance on the Train data. But the neural network performance degrades in the Validate data set but seems to recover slightly in the Test data set.

The top 10% performance is closest in the Test data, with the decision tree showing the best lift. The logistic regression starts to perform better at 40% of the file and higher.

Choosing the best model presents an interesting challenge. Since they are very close, you could select the decision tree model because it is the easiest to interpret and offers the greatest lift in the top 10%. Since our loan default rate is approximately 4%, it makes sense to select a model that optimizes in the top 10%. On the other hand, if the loan default rate was 40%, you might select the Stepwise Logistic Regression model, since it performs better at 40% of the file.

**Output 9.13: Model Comparison—Cumulative Lift**

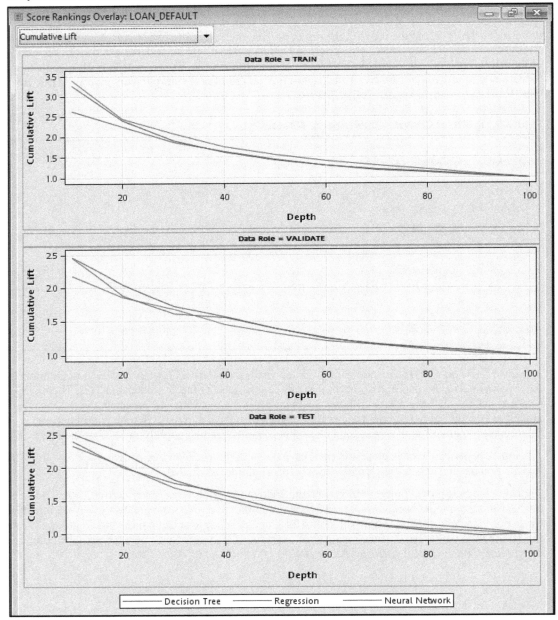

Next, you will look at the gains tables. With the cumulative lift section (lower left) selected, go to the top left menu and click the **Table** icon (second icon from the right) as shown in Figure 9.18.

**Figure 9.18: Model Comparison Property Window**

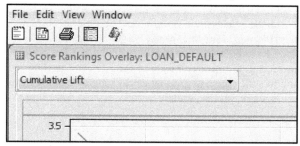

In Output 9.14, you see a partial view of the gains tables for all three models for all three data sets. Because we selected 10 bins in the Property window, the values are in deciles, as were those shown in Figure 9.13.

**Output 9.14: Gains Tables**

Table: Score Rankings Overlay: LOAN_DEFAULT

| Predecessor Node | Model Node | Model Description | Data Role | Target Variable | Target Label | Event | Cumulative % Response |
|---|---|---|---|---|---|---|---|
| Tree | Tree | Decision Tr... | TRAIN | LOAN_DEF... | | 1 | 13.23926 |
| Tree | Tree | Decision Tr... | TRAIN | LOAN_DEF... | | 1 | 9.752745 |
| Tree | Tree | Decision Tr... | TRAIN | LOAN_DEF... | | 1 | 7.696946 |
| Tree | Tree | Decision Tr... | TRAIN | LOAN_DEF... | | 1 | 6.593571 |
| Tree | Tree | Decision Tr... | TRAIN | LOAN_DEF... | | 1 | 5.796509 |
| Tree | Tree | Decision Tr... | TRAIN | LOAN_DEF... | | 1 | 5.228133 |
| Tree | Tree | Decision Tr... | TRAIN | LOAN_DEF... | | 1 | 4.822565 |
| Tree | Tree | Decision Tr... | TRAIN | LOAN_DEF... | | 1 | 4.518087 |
| Tree | Tree | Decision Tr... | TRAIN | LOAN_DEF... | | 1 | 4.28151 |
| Tree | Tree | Decision Tr... | TRAIN | LOAN_DEF... | | 1 | 4.05217 |
| Tree | Tree | Decision Tr... | VALIDATE | LOAN_DEF... | | 1 | 10.59936 |
| Tree | Tree | Decision Tr... | VALIDATE | LOAN_DEF... | | 1 | 8.782934 |
| Tree | Tree | Decision Tr... | VALIDATE | LOAN_DEF... | | 1 | 7.428662 |
| Tree | Tree | Decision Tr... | VALIDATE | LOAN_DEF... | | 1 | 6.7179 |
| Tree | Tree | Decision Tr... | VALIDATE | LOAN_DEF... | | 1 | 5.988182 |
| Tree | Tree | Decision Tr... | VALIDATE | LOAN_DEF... | | 1 | 5.401544 |
| Tree | Tree | Decision Tr... | VALIDATE | LOAN_DEF... | | 1 | 4.983028 |
| Tree | Tree | Decision Tr... | VALIDATE | LOAN_DEF... | | 1 | 4.668686 |
| Tree | Tree | Decision Tr... | VALIDATE | LOAN_DEF... | | 1 | 4.424219 |
| Tree | Tree | Decision Tr... | VALIDATE | LOAN_DEF... | | 1 | 4.300788 |
| Tree | Tree | Decision Tr... | TEST | LOAN_DEF... | | 1 | 9.988095 |

If you scroll through, you can select two sections to compare. First, click the column heading called **Data Role** to sort the data by data role and put all the test data in contiguous rows. For the **Test** data set, select the results from the decision tree and the logistic regression. Within the Table Score Rankings window, you can rearrange the columns together that you want to capture. Slide the **Baseline Cumulative % Captured Response**, **Number of Events (LOAN_DEFAULT)**, **Mean Posterior Probability**, **Cumulative % Captured Response**, and **Cumulative Lift** to the left, next to the column **Data Role**. Click the **Data Role** column heading to order the rows alphabetically by data role. Doing so puts the two tables of interest, the decision tree test data and the logistic regression test data, into the top 20 rows. The result is captured in Output 9.15.

**Output 9.15: Gains Tables Column Headings**

| Model Description | Data Role ▲ | Baseline Cumulative % Captured Response | Number of Events | Mean Posterior Probability | Cumulative % Captured Response | Cumulative Lift |
|---|---|---|---|---|---|---|
| Decision Tr... | TEST | 10 | 74.91071 | 0.132006 | 25.30767 | 2.528405 |
| Decision Tr... | TEST | 20 | 57.03929 | 0.062442 | 44.5777 | 2.22829 |
| Decision Tr... | TEST | 30 | 28.84646 | 0.034881 | 54.32313 | 1.81069 |
| Decision Tr... | TEST | 40 | 25.52757 | 0.032812 | 62.94731 | 1.573263 |
| Decision Tr... | TEST | 50 | 17.03925 | 0.025822 | 68.70381 | 1.373893 |
| Decision Tr... | TEST | 60 | 14.52405 | 0.02388 | 73.61059 | 1.226789 |
| Decision Tr... | TEST | 70 | 14.54345 | 0.02388 | 78.52391 | 1.121578 |
| Decision Tr... | TEST | 80 | 14.52405 | 0.02388 | 83.43069 | 1.042779 |
| Decision Tr... | TEST | 90 | 14.52405 | 0.02388 | 88.33746 | 0.981484 |
| Decision Tr... | TEST | 100 | 34.52111 | 0.019751 | 100 | 1 |
| Regression | TEST | 10 | 71.40351 | 0.107348 | 24.12281 | 2.410029 |
| Regression | TEST | 20 | 47.41739 | 0.06226 | 40.14219 | 2.006574 |
| Regression | TEST | 30 | 38.07384 | 0.048572 | 53.00498 | 1.766754 |
| Regression | TEST | 40 | 35.36587 | 0.038908 | 64.95291 | 1.623389 |
| Regression | TEST | 50 | 29.7229 | 0.03092 | 74.99443 | 1.499688 |
| Regression | TEST | 60 | 14.92065 | 0.027071 | 80.03519 | 1.33386 |
| Regression | TEST | 70 | 15.77017 | 0.021753 | 85.36295 | 1.219261 |
| Regression | TEST | 80 | 15.74914 | 0.021753 | 90.68361 | 1.133432 |
| Regression | TEST | 90 | 16.45338 | 0.021358 | 96.24218 | 1.06931 |
| Regression | TEST | 100 | 11.12315 | 0.015999 | 100 | 1 |

By comparing these two tables, you can see that the results are very similar. When comparing **Cumulative Lift**, you see that the decision tree out-performs the logistic regression model in each of the top three deciles. Consequently, according to this performance, the decision tree model is the best predictor of loan default.

The **Cumulative Lift** can be thought of as the amount of improvement over random at any given depth of file. So, it's roughly the **Cumulative % of Captured Response** divided by the **Baseline Cumulative % of Captured Response** or the expected result without using a model. In the first decile (top 10%), the model predicts loan default roughly 2.5 times more accurately than without a model. With expected losses in the millions, the ability to avoid potentially bad loans can have a huge benefit to the lender's profits.

A common strategy for risk models is to assign rules based on the ranges of probability. For example, you can automatically approve any application with a mean probability of default that is less than a specified value. If you look at the **Mean Posterior Probability** for the decision tree model, you might say that any loan applicant with a probability of default less than 3% is automatically approved. Any loan applicant with a probability of default greater than 4% is automatically declined, and those applicants with a probability of default between 3% and 4% require further research.

The benefits of predictive models have proven to save or make companies millions of dollars. For more information about applying models, see *Data Mining Cookbook* (Parr-Rud 2001). For information about common modeling issues and mistakes, see *Identifying and Overcoming Common Data Mining Mistakes* (Wielenga 2007).

> **NOTE:** If you are tasked with building risk models, SAS Enterprise Miner offers an advanced feature that allows you to build scorecards. For more information, see your SAS representative.

## Notes from the Field

By mastering the techniques in this chapter, you have a powerful set of skills you can use to deliver strategic and tactical business analysis to your clients and customers. With this power comes great responsibility. Keep in mind that your clients may not understand how you get your results, or even know if they are accurate. So, you must practice due diligence. If you are on a team, perhaps you can review your results with another analyst. Or, review your results with someone who has deep business knowledge in the subject matter of your analysis.

When presenting the results of your analysis, always tie your findings back to your business objectives. Explain to your stakeholders in clear business language how they can take action or derive insights from the results. Avoid statistical jargon such as *chi-squared* and *log likelihood*. They will understand lift, especially if you can tie it to increased savings or decreased loss.

Now, you can consider your Ferrari to be fully loaded; it has plenty of power under its hood, and it has many features that will enable you to drive through difficult terrain and reach your destination. You may not understand how the engine or transmission works. You may not have a clue about the suspension or electrical systems. But, you have gauges that can alert you to your progress, and warnings when things require adjustment. Most of all, you have reason to enjoy the ride.

# References

Association of Certified Fraud Examiners. 2012. *Report to the Nations on Occupational Fraud and Abuse: 2012 Global Fraud Study.* http://www.acfe.com/uploadedFiles/ACFE_Website/Content/rttn/2012-report-to-nations.pdf.

Cody, Ron. 2008. *Cody's Data Cleaning Techniques Using SAS*, 2nd ed. Cary, NC: SAS Institute Inc.

Delwiche, Lora D., and Susan J. Slaughter. 2010. *The Little SAS Book for Enterprise Guide 4.2.* Cary, NC: SAS Institute Inc.

DeVille, Barry, and Padraic Neville. 2013. *Decision Trees for Analytics Using SAS Enterprise Miner.* Cary, NC: SAS Institute Inc.

Parr-Rud, Olivia. 2001. *Data Mining Cookbook: Modeling Data for Marketing, Risk, and Customer Relationship Management.* New York: Wiley.

SAS Institute Inc. 2014. "Big Data: What It Is and Why It Matters." SAS Insights Center. http://www.sas.com/en_us/insights/big-data/what-is-big-data.html.

Siegel, Eric, and Thomas H. Davenport. 2013. *Predictive Analytics: The Power to Predict Who Will Click, Buy, Lie, or Die.* New York: Wiley.

Wielenga, Doug. 2007. *Identifying and Overcoming Common Data Mining Mistakes.* SAS Global Forum Paper 073-2007. Cary, NC: SAS Institute Inc. http://www2.sas.com/proceedings/forum2007/073-2007.pdf.

# Index

## A

## B

## C

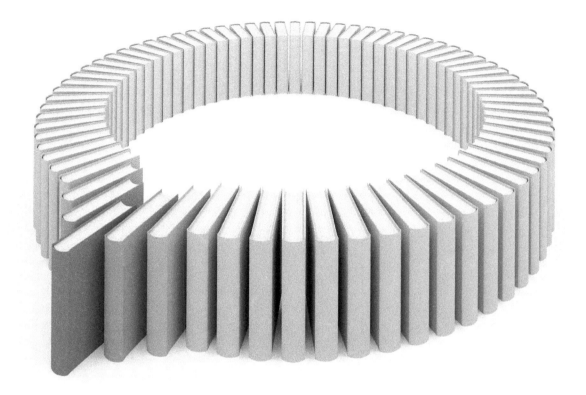

# Gain Greater Insight into Your SAS® Software with SAS Books.

Discover all that you need on your journey to knowledge and empowerment.

support.sas.com/bookstore
*for additional books and resources.*